Semi-Active Suspension Control Design for Vehicles

Semi-Active Suspension Control Design for Vehicles

S.M. Savaresi
C. Poussot-Vassal
C. Spelta
O. Sename
L. Dugard

AMSTERDAM • BOSTON • HEIDELBERG • LONDON • NEW YORK • OXFORD
PARIS • SAN DIEGO • SAN FRANCISCO • SINGAPORE • SYDNEY • TOKYO
Butterworth-Heinemann is an imprint of Elsevier

Butterworth-Heinemann is an imprint of Elsevier
The Boulevard, Langford Lane, Kidlington, Oxford, OX5 1GB, UK
30 Corporate Drive, Suite 400, Burlington, MA 01803, USA

First published 2010

British Library Cataloguing in Publication Data
Semi-active suspension control design for vehicles.
1. Active automotive suspensions–Design.
I. Savaresi, Sergio M.
629.2'43–dc22

Library of Congress Control Number: 2010925093

ISBN: 978-0-08-096678-6

For information on all Butterworth-Heinemann publications
visit our Website at *www.elsevierdirect.com*

Typeset by: diacriTech, India

Printed and bound in China
10 11 12   11 10 9 8 7 6 5 4 3 2 1

Dedication

To Cristina, Claudio and Stefano (S.M.S)

To my Family (C.P-V)

To Daniela (C.S.)

To Isabelle, Corentin and Grégoire (O.S.)

To Brigitte (L.D.)

Contents

List of Figures

List of Tables

About the Authors

Sergio Matteo Savaresi was born in Manerbio, Italy, in 1968. He received an M.Sc. in Electrical Engineering (Politecnico di Milano, 1992), a Ph.D. in Systems and Control Engineering (Politecnico di Milano, 1996), and an M.Sc. in Applied Mathematics (Catholic University, Brescia, 2000). After the Ph.D. he worked as management consultant at McKinsey & Co, Milan Office. He has been Full Professor in Automatic Control at Politecnico di Milano since 2006, and head of the "mOve" research team (http://move.dei.polimi.it/). He was visiting researcher at Lund University, Sweden; University of Twente, The Netherlands; Canberra National University, Australia; Stanford University, USA; Minnesota University at Minneapolis, USA; and Johannes Kepler University, Linz, Austria. He is Associate Editor of: the *IEEE Transactions on Control System Technology*, the *European Journal of Control*, the *IET Transactions on Control Theory and Applications*, and the *International Journal of Vehicle Systems Modelling and Testing*. He is also Member of the Editorial Board of the IEEE CSS. He is author of more than 250 scientific publications at international level (involving many patents), and he has been the proposer and manager of more than 50 sponsored joint research projects between the Politecnico di Milano and private companies. His main interests are in the areas of vehicles control, automotive systems, data analysis and system identification, nonlinear control theory, and control applications. He is married to Cristina and has two sons, Claudio and Stefano.

Charles Poussot-Vassal was born in Grenoble, France, in 1982. In 2005, he completed his Engineering degree and M.Sc. in Control and Embedded Systems from Grenoble INP-ESISAR (Valence, France) and Lund University of Technology (Lund, Sweden), respectively. In 2008, he completed his Ph.D. degree in Control Systems Theory, with applications of linear parameter varying modeling and robust control methods on automotive systems (suspension and global chassis control) at the GIPSA-lab's control systems department, from the Grenoble Institute of Technology (Grenoble, France), under the supervision of O. Sename and L. Dugard. He has been a visiting student with the MTA SZTAKI, University of Budapest (Budapest, Hungary), under the supervision of J. Bokor, P. Gáspár and Z. Szabó. At the beginning of 2009, he worked as a Research Assistant with the Politecnico di Milano (Milan, Italy) on semi-active suspension control, under the supervision of S.M. Savaresi. From mid-2009, he has been Researcher with ONERA, the French aerospace lab, with the Flight Dynamics and Control Systems department. His main interests concern control system design, model reduction techniques and dynamical performance analysis, with application in ground vehicles, web servers and aircraft systems.

Cristiano Spelta was born in Milan, Italy, on 20 March 1979. He received a Masters degree in Computer Engineering in 2004 from the Politecnico di Milano. He earned from the same university a Ph.D. in Information Engineering in 2008 (thesis "Design and applications of semi-active control systems"). He was visiting scholar (July–September 2006) at the Institute of Control Sciences of Moscow under the supervision of Professor Boris Polyak. He is currently Assistant Professor at the Università degli Studi di Bergamo (BG, Italy). He is author of more than 30 international publications including some industrial patents. His research interests include control of road and rail vehicles, control problems in system integration, and robust control and mixed $\mathcal{H}_2$-$\mathcal{H}_\infty$ control problems.

Olivier Sename received a Ph.D. degree in 1994 from the Ecole Centrale Nantes, France. He is now Professor at the Grenoble Institute of Technology (Grenoble INP), within the GIPSA-lab. His main research interests include theoretical studies in the field of time-delay systems, linear parameter varying systems and control/real-time scheduling co-design, as well as robust control for various applications such as vehicle dynamics, engine control. He has collaborated with several industrial partners (Renault, SOBEN, Delphi Diesel Systems, Saint-Gobain Vetrotex, PSA Peugeot-Citroën, ST Microelectronics), and is responsible for international bilateral research projects (Mexico, Hungary). He is the (co-)author of 6 book chapters, 20 international journal papers, and more than 80 international conference papers. He has supervised 15 Ph.D. students.

Luc Dugard works as a CNRS Senior Researcher (Directeur de Recherche CNRS) in the Automatic Control Dept. of GIPSA-lab, a research department of Grenoble INP (Institut Polytechnique de Grenoble), associated to the French research organization "Centre National de la Recherche Scientifique". Luc Dugard has published about 90 papers and/or chapters in international journals or books and more than 220 international conference papers. He has co-advised 28 Ph.D. students. His main research interests include (or have included) theoretical studies in the field of adaptive control, robust control, and time delay systems. The main control applications are oriented towards electromechanical systems, process control and automotive systems (suspensions, chassis and common rail systems).

Preface

The suspension (together with the tire), is probably the single element of a vehicle which mostly affects its entire dynamic behavior. It is not surprising that in the most essential and fun-driving vehicles – e.g. sport motorbikes – suspensions play a central role (sometimes almost "worshipped" by their owner) with an intriguing mixture of technical features and aesthetic appeal.

This central role of suspensions in vehicle dynamics is intuitive: they establish the link between the road and the vehicle body, managing not only the vertical dynamics, but also the rotational dynamics (roll, pitch) caused by their unsynchronized motions. As such, they contribute to create most of the "feeling" of the vehicle, affecting both its safety and driving fun.

Another peculiar feature of the suspensions in a vehicle is their possible appearance at different layers: at the classical wheel-to-chassis layer, at the chassis-to-cabin layer (e.g. in trucks, earth-moving machines, agricultural tractors, etc.) and at the cabin-to-seat layer (in large vehicles with suspended cabins the driver seat is also typically equipped with a fully fledged suspension system).

The Italian cartingent of the authors of this book would be loathe to admit it, but the birth of electronic suspensions for the car mass-market can probably be dated back to the early 1960s, when Citroën introduced hydro-pneumatic suspensions in its top cars. At that time those suspensions were still untouched by electronics (they were "ante-litteram" electronic suspensions), but the idea of having part of a suspension so dramatically and easily modified opened the way to the idea of "on-line" electronic adaptation of the suspension.

Given this tribute to Monsieur Citroën, the real "golden age" of electronic suspension can be probably located in the 1980s; analog electronics were already well-developed, the era of embedded digital micro-controllers was starting, and the magic of fully active suspensions attracted both the F1 competitions and the car manufacturers. During these years the exceptional potential of replacing a traditional spring-damper system with a fully fledged electronically controllable fast-reacting hydraulic actuator was demonstrated.

High costs, significant power absorption, bulky and unreliable hydraulic systems, uncertain management of the safety issues: the fatal attraction for fully active electronic suspensions lasted only a few years. They were banned by F1 competitions in the early 1990s and they have never had (so far) a significant impact on mass-market car production.

In the second half of the 1990s, a new trend emerged: it became increasingly clear that the best compromise of cost (component cost, weight, electronics and sensors, power consumption, etc.) and performance (comfort, handling, safety) was to be found in another technology of electronically controllable suspensions: the variable-damping suspension or, in brief, the semi-active suspension.

After a decade this technology is still the most promising and attractive: it has been introduced in the mass-market production of cars; it is entering the motorcycle market; a lot of special vehicles or niche applications are considering this technology; many new variable-damping technologies are being developed.

Semi-active suspensions are expected to play an even more important role in the new emerging trend of electric vehicles with in-wheel motors: in such vehicle architecture the role of suspension damping is more crucial, and semi-active suspensions can significantly contribute to reduce the negative effects of the large unsprung mass.

The scope of this book is to present a complete discussion of the problem of designing control algorithms for semi-active suspensions. Even though the effect of a modification of the damping coefficient of a suspension is well-known, when damping-coefficient variation is carried out at a very fast rate (e.g. every 5 milliseconds), making a decision on the "best" damping ratio is far from easy.

A semi-active suspension system is an unusual combination of seemingly simple dynamics (whose bulk can be easily captured by a fourth-order model) and challenging features (nonlinear behavior, time-varying parameters, asymmetrical control bounds, uncontrollability at steady-state, etc.). These features make the design of semi-active control algorithms very challenging. This gives the opportunity, by "simply" changing the control strategy, to modify significantly the dynamic behavior of a vehicle. However, this is an opportunity which is not easy to catch: the history of semi-active suspensions is full of anecdotes about semi-active suspensions being rejected by vehicle manufacturers just because they "do not make any difference...", or even "are worse than the (nice, old) traditional mechanical suspensions...". As in many other electronically controlled systems, the actuator is not "smart itself": it simply inherits the smartness (or dumbness) of its control-algorithm designer.

The key of semi-active suspensions is in the algorithm. The design of semi-active control algorithms is the aim of this book.

The structure of the book follows the classical path of the control-system design: first, the actuator (the variable-damping shock absorber) is discussed, modeled, and the available technologies are presented. Then the vehicle (equipped with semi-active dampers) is mathematically modeled, and the control algorithms are designed and discussed.

This book can be effectively accessed at three reading levels: a tutorial level for students; an application-oriented level for engineers and practitioners; and a methodology-oriented level for researchers. To enforce these different reading levels, and to present the material in an incremental manner from the basic to the most advanced control approaches, the book has been conceptually divided into two parts.

In the first part of the book, made up of Chapters 2 to 6, where the basics of modeling and semi-active control design are described, whereas in the second part of the book, made up of Chapters 6 to 8, more advances and research-oriented solutions are proposed and compared, with the help of some case studies. Overall, the first part of the book presents the topic at a level of depth which can be considered appropriate for practitioners and for a course on vehicle control at the M.Sc. level, while the second part constitutes additional material of interest for graduate studies and for researchers in automotive control.

It is also worth noting that Chapter 4 ("Methodology of analysis for automotive suspensions") and Chapter 5 ("Optimal strategy for semi-active suspensions and benchmark") play a pivotal role in the organization of the book:

- In Chapter 4, the different techniques and methods for evaluating the performance of a suspension system are discussed in detail, in order to have a common baseline to assess and compare the quality of different design solutions.
- In Chapter 5 an "ideal" semi-active control strategy is developed, by assuming full knowledge (past and future) of the road profile, and using a sophisticated off-line numerical optimization based on model-predictive control. Even though this control strategy cannot be implemented in practice, it is conceptually very important since it sets an absolute bound for the best possible filtering performance of semi-active suspensions, and represents a simple and clear benchmark for any "real" algorithm.

It is also worth noticing that most of the material presented in the book focuses on vertical dynamics only: it constitutes the bulk of suspension control, and most of the pitch and roll control-design problems are inherently solved by applying the semi-active control strategy to each corner of the vehicle, or solutions can be straightforwardly derived from the vertical-dynamics algorithms.

Finally, a few words on the unusual author team. Despite the (comparatively) long list of authors and their different affiliations, this book is not an "edited" book, made up from an inhomogeneous collection of different contributions, but it is the result of a real effort to condense in an instructive way most of the main results and research work which has been developed in the last decade on this topic.

This book incorporates all the research work and the cooperation with suspension and vehicle manufacturers that Politecnico di Milano and Grenoble University have accumulated on this topic, obtaining, we hope, the best of both experiences.

The composition of the author team also proves that Italy and France can continue their long-lasting tradition of stimulating and successful cooperation . . . even after the '06 Berlin World Championship final.

Milano and Grenoble, January 15, 2010

Sergio Matteo Savaresi
Charles Poussot-Vassal
Cristiano Spelta
Olivier Sename
Luc Dugard

Acknowledgements

Italian Authors

This book is the result of several years of collaboration both with academic and industrial partners.

We are grateful to all our co-authors of the papers we have written in the preceding years on the topic of electronic suspensions: Sergio Bittanti, Fabio Codecá, Diego Delvecchio, Daniel Fischer, Rolf Isermann, Lorenzo Nardo, Enrico Silani, Francesco Taroni, Simone Tognetti, Simone Tremolada.

In the industrial world, we are particularly indebted to Luca Fabbri, Mario Santucci, Lorenzo Nardo and Onorino di Tanna of Piaggio Group S.p.A., Sebastiano Campo, Andrea Fortina, Fabio Ghirardo, Gabriele Bonaccorso and Andrea Moneta of FIAT Automobiles S.p.A., Mauro Montiglio of Centro Ricerche FIAT, Andrea Stefanini of Magneti Marelli, Joachim Funke of Fludicon Gmbh, Kristopher Burson of LORD Corp., Lars Jansson and Henrik Johansson of Öhlins Racing AB, Piero Vicendone of ZF Sachs Italia S.p.A, Gianni Mardollo of Bitubo, Andrea Pezzi of Marzocchi-Tenneco, Riccardo and Andrea Gnudi of Paioli Meccanica, Ivo Boniolo of E-Shock, Filippo Tosi of Ducati Corse, and Fabrizio Palazzo of Yamaha Motorsport Europe, for their constant support and interest in investigating advanced solutions and for providing us with an industrial perspective on several research topics. Special thanks to Vittore Cossalter, a passionate motorcyclist and great expert of suspension mechanics.

The material presented in this book has also been developed thanks to the activity of the MOtor VEhicle control team (http://move.dei.polimi.it/) of the Politecnico di Milano; we would like to thank all its present and past members for their collaboration over the years.

Further, we want to thank all our present and former students, who helped us to organize and refine the presentation of the different topics since the beginning of the course on Vehicle Control at the Politecnico di Milano.

French Authors

We would like to thank first the former and present students (in particular the Ph.D. students) who have worked on suspension systems and have been co-authors of the referenced papers: Marek Nawarecki, Damien Sammier, Carsten Lueders, Alessandro Zin, Sébastien Aubouet, Anh-Lam Do and Jorge Lozoya.

We also are grateful to our partners abroad we are collaborating with, leading to an extension of our knowledges and skills in that field: Ricardo Ramirez-Mendoza, Ruben Morales, Aline Drivet and Leonardo Flores (Tecnologico de Monterrey, Mexico), Peter Gáspár, Zoltan Szabó and József Bokor (University of Budapest, Hungary) and Michel Basset (Université de Haute Alsace, France).

Finally, the industrial collaboration with PSA Peugeot-Citroën (Vincent Abadie and Franck Guillemard) launched us on semi-active suspension control. This is now continuing with SOBEN (Benjamin Talon). We would like to specifically thank these people.

Notations

Table 1: List of mathematical symbols and variables used in the book

Mathematical notation	Meaning
$\mathbb{R}$	Real values set
$\mathbb{R}^+$ ($\mathbb{R}^{+*}$)	Positive real values set (without 0)
$\mathbb{C}$	Complex values set
$\mathbb{C}^+$ ($\mathbb{C}^{+*}$)	Positive complex values set (without 0)
A^T	Transpose of $M \in \mathbb{R}$
A^*	Conjugate of $M \in \mathbb{C}$
$A + (\star)^T = A + A^T$	Defines the transpose matrix of $A \in \mathbb{R}$
$A + (\star)^* = A + A^*$	Defines the conjugate matrix of $A \in \mathbb{C}$
$A = A^T$	Matrix A is real symmetric
$A = A^*$	Matrix A is hermitian
$M \prec (\preceq)0$	Matrix M is symmetric and negative (semi)definite
$M \succ (\succeq)0$	Matrix M is symmetric and positive (semi)definite
$\mathbf{Tr}(A)$	Trace of A matrix (sum of the diagonal elements)
$\mathbf{Co}(A)$	Convex hull of set A
$\sigma(.)$	Singular value ($\sigma(A)$ defines the eigenvalues of the operator $(A^*A)^{1/2}$)
$Re(.)$	Real part of a complex number
$Im(.)$	Imaginary part of a complex number
j	Complex variable
s	Laplace variable $s = j\omega$, where ω is the pulsation
$\omega = 2\pi f$	Pulsation in rad/s
$\dot{x} = \frac{d}{dt}x(t)$	Derivative of function $x(t)$ with respect to t
$\int x(t)dt$	Integral of function $x(t)$ with respect to t
$\sum_i x_i$	Sum of the x_i elements

Table 2: List of acronyms used in the book

Acronyms	Meaning
BMI	Bilinear matrix inequality
LMI(s)	Linear matrix inequality(ies)
LTI	Linear time invariant
LTV	Linear time variant
LPV	Linear parameter varying
qLPV	Quasi linear parameter varying
SDP	Semi-definite programming
ABC	Active body control
ABS	Anti-locking braking system
COG	Center of gravity
DOF	Degree of freedom
ERD	Electrorheological damper
EHD	Electrohydrological damper
MRD	Magnetorheological damper
SER	Speed effort rule (force provided by the damper as a function of the deflection velocity)
ADD	Acceleration driven damper
GH	Ground-Hook (or Groundhook)
LQ	Linear quadratic
SH	Sky-Hook (or Skyhook)
MPC	Model predictive control
PDD	Power driven damper
iff.	if and only if
s.t.	such that/so that
resp.	respectively
w.r.t.	with respect to

Table 3: List of model variables used in the book (unless explicitly specified)

Variable	Meaning
M	Suspended mass
m	Unsprung (tire) mass
k	Stiffness coefficient
c	Damping coefficient
c_{min} (c_{max})	Minimal (maximal) damping coefficient
k_t	Tire stiffness coefficient
c_t	Tire damping coefficient
g	Gravitational constant
L	Nominal suspension length
R	Nominal tire radius
I_x	x-axis inertia
I_y	y-axis inertia
I_w	Wheel inertia
β	Actuator bandwith
μ	Tire/road adhesion coefficient
l	Vehicle length
t	Vehicle width
F_k	Suspension stiffness force
F_d	Suspension damping force
$F_{sz} = F_k + F_d$	Suspension force
F_{kt}	Tire stiffness force
F_{dt}	Tire damping force
F_{tx}	Tire longitudinal force
F_{ty}	Tire lateral force
$F_{tz} = F_{kt} + F_{dt}$	Tire vertical force
F_L	Vertical force load
T_b	Wheel braking torque
v	Vehicle velocity at the COG
λ	Wheel slip ratio
ω	Wheel rotational velocity
x	Vehicle longitudinal displacement
y	Vehicle lateral displacement
z	Chassis vertical displacement
z_t	Wheel vertical displacement
ϕ	Vehicle pitch angle
θ	Vehicle roll angle

CHAPTER 1

Introduction and Motivations

1.1 Introduction and Historical Perspective

A suspension, in its more classical and conventional configuration (see Figure 1.1) is constituted by three main elements:

- An elastic element (typically a coil spring), which delivers a force proportional and opposite to the suspension elongation; this part carries all the static load.
- A damping element (typically a hydraulic shock absorber), which delivers a dissipative force proportional and opposite to the elongation speed; this part delivers a negligible force at steady-state, but plays a crucial role in the dynamic behavior of the suspension.
- A set of mechanical elements which links the suspended (sprung) body to the unsprung mass.

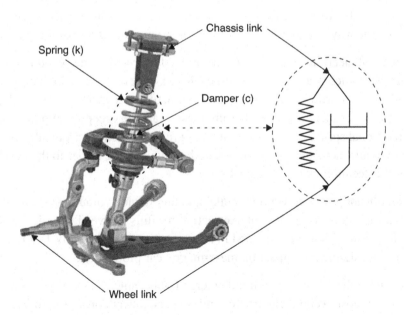

Spring (k)

Chassis link

Damper (c)

Wheel link

Figure 1.1: Classical scheme of a wheel-to-chassis suspension in a car.

DOI: 10.1016/B978-0-08-096678-6.00001-8

From the dynamic point of view, the spring and the damper are the two key elements, whereas the mechanical links are mainly responsible to the suspension kinematics; hence, in the rest of the book, the focus will be on the two "dynamic" elements of the suspension.

Roughly speaking, the suspension is a mechanical low-pass filter which attenuates the effects of a disturbance (e.g. an irregular road profile) on an output variable. The output variable is typically the body acceleration when comfort is the main objective; the tire deflection when the design goal is road-holding. From Figure 1.2 it is clear that these two objectives are somehow conflicting: the tuning and the design of a mechanical suspension tries to find the best compromise between these two goals.

This critical trade-off is worsened by the fact that a suspension has a limited travel; when the end-stop (bushing) of a suspension is reached, both the comfort and road-holding performances are dramatically deteriorated, and the occurrence of this situation must be carefully avoided.

All in all, the bulk of the design problem of a classical mechanical suspension consists in the definition of a spring stiffness and a damping ratio, in order to deliver a good compromise between comfort and handling, with an additional bound on the suspension travel. Given such a tricky set of trade-offs, it is not surprising that, when the early age of car manufacturing ended, suspension designers began to look for possible ways to reduce the problem of compromising between opposite goals. In this respect, the birth of electronic suspensions for the car mass-market can probably be dated from the early 1960s, when Citroën introduced hydro-pneumatic suspensions (Figure 1.3). At this time suspensions were still completely electronics-free, but the idea of having part of a suspension so dramatically and easily modified opened the way to the idea of "on-line" electronic adaptation of the suspension.

The "golden age" of electronic suspension was probably located in the second half of the 1980s; analog electronics were already well-developed, the era of embedded digital micro-controllers was starting, and the magic of full-active suspensions attracted both the F1 competition and the car manufacturers. During these years the exceptional potential of replacing a traditional spring-damper system with a fully-fledged electronically controllable fast-reacting hydraulic actuator was demonstrated. Lotus was the leader in the development and testing of this technology (see Figure 1.4).

High costs, significant power absorption, bulky and unreliable hydraulic systems, uncertain management of the safety issues: the fatal attraction for fully-active electronic suspensions lasted only a few years. They were banned in F1 competitions in the early 1990s, and they never had (so far) a significant impact on mass-market car production.

In the second half of the 1990s, a new trend emerged: it became clear that the best compromise of cost (component cost, weight, electronics and sensors, power consumption, etc.) and

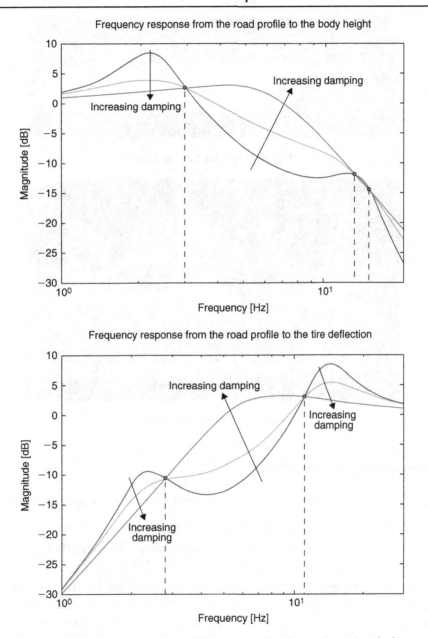

Figure 1.2: Filtering effect of a passive suspension: example of a road-to-chassis frequency response (up), and a road-to-tire-deflection frequency response (bottom).

performance (comfort, handling, safety) lay in another technology of electronically controllable suspensions, namely, the variable-damping suspensions or, in brief, the semi-active suspensions.

Figure 1.3: The Citroën DS.

Figure 1.4: The Lotus Excel.

1.2 Semi-Active Suspensions

Electronically controlled suspensions can be classified according to two main features:

- **Energy input**: when energy is "added" into the suspension system, the suspension is classified as "active"; however, when the suspension is electronically modified without energy insertion (apart from a small amount of energy used to drive the electronically controlled element), the suspension is called semi-active. Roughly speaking, a suspension is "active" when it can "lift" the vehicle, semi-active otherwise.
- **Bandwidth**: the electronically controlled element of the suspension can be modified with a specific reaction-time; this feature strongly characterizes the suspension system, since it inherently defines the maximum achievable bandwidth of the corresponding closed-loop control system.

According to the above two features, there are five main classes of electronically controlled suspensions (see e.g. Guglielmino et al., 2008; Hrovat, 1997; Isermann, 2003):

- Three "active" suspensions:

  - Load-leveling suspensions (namely active suspensions with an actuation bandwidth well below the main suspension dynamics).
  - Slow-active suspensions (active suspensions with a bandwidth in between body and wheel dynamics).
  - Fully-active suspensions (full-bandwidth active suspensions).
- Two "semi-active" suspensions:
  - Adaptive suspensions (suspension with slowly-modified damping ratio; typically this modification is simply made with an open-loop architecture).
  - Semi-active suspensions (suspensions with a damping ratio modified in a closed-loop configuration over a large bandwidth).

Today the most appealing electronic-suspension configuration is constituted by the combination of load-leveling systems (e.g. with a gas spring) and semi-active dampers (see e.g. Figure 1.5). Notice that, from the control design point of view, the load-leveling part of the suspension is rather trivial, whereas the design of the semi-active part is very challenging.

Semi-active suspensions are an amazing mix of appealing features; among others, the most interesting are the following:

- **Negligible power-demand**: since they are based on the regulation of the damping ratio only, the power-absorption is limited to a few Watts required to modify the hydraulic orifices or the fluid viscosity.

Figure 1.5: Example of a suspension of a luxury sedan (Audi A8), which integrates an electronically controlled gas spring with load-leveling capabilities, and a semi-active damper.

- **Safety**: in a semi-active suspension the stability is always guaranteed by the fact that the whole system remains dissipative, whatever the damping ratio is.
- **Low cost, low weight**: the main damping-modulation technologies (electrohydraulic, magnetorheological, electrorheological, air-damping) can be produced (for large volumes) at low cost and with compact packagings.
- **Significant impact on the vehicle performance**: by changing the damping ratio of a suspension the overall comfort and road-holding performance can be significantly modified.

Changing the damping ratio represents a very interesting opportunity for a suspension designer; however, the selection of the best damping ratio is not an easy task, even in the simple case when the damping ratio is not subject to fast-switching (see Figure 1.6). The task becomes extremely challenging when the suspension designer has the opportunity to change (possibly with a feedback control scheme, using vehicle-dynamics sensors like accelerometers and potentiometers) the damping ratio every (e.g.) 5 milliseconds. In this case the real "key problem" becomes the control-algorithm design problem.

The potential benefit of sophisticated control algorithms applied to a fast-switching electronic damper can be easily appreciated in Figure 1.7, where the filtering (comfort-oriented) performance of three different semi-active control algorithms: (labeled SH-C, Mix-1, and

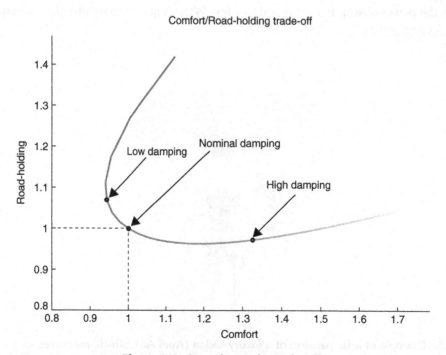

Figure 1.6: Damping-ratio trade-off.

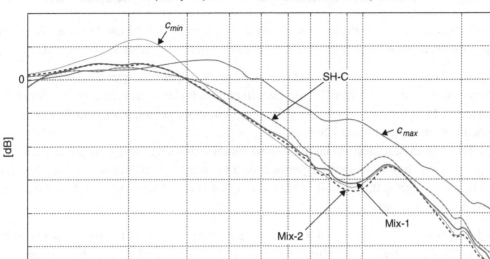

Figure 1.7: An experimental comparison of filtering performance (comfort objective): semi-active strategies; labeled SH-C (for Skyhook), Mix-1 (for Mixed Skyhook-ADD with 1 sensor) and Mix-2 (for Mixed Skyhook-ADD with 2 sensors) versus fixed-damping configurations (c_{min} and c_{max}).

Mix-2; they will be presented in detail in the second part of the book – see Ahmadian et al., 2004; Karnopp et al., 1974; Savaresi and Spelta, 2009) are compared with the performance of fixed-damping configuration (labeled as c_{min} and c_{max}, corresponding to a low-damping and a high-damping configuration, respectively). From Figure 1.7 two main conclusions can be easily drawn:

- The fixed-damping configurations have an intrinsic trade-off: a low-damping provides superior high-frequency filtering performance, but it is affected by a badly undamped body resonance; on the other hand, a high-damping setting removes the resonances, but strongly deteriorates the filtering capabilities. Intermediate damping settings simply deliver different combinations of this trade-off.
- A wise semi-active algorithm can (almost) completely remove the classical trade-off: good damping of the body resonance can be guaranteed, together with good filtering performance.

The aim and scope of this book is to enter into the challenging problem of designing semi-active control algorithms. More than in other vehicle control applications, in this case it is the algorithm which makes the difference.

1.3 Applications and Technologies of Semi-Active Suspensions

Thanks to their appealing features, today semi-active suspensions are used over a vast domain of applications. In vehicle applications, semi-active suspensions are used at different layers:

- At the (classical) wheel-to-chassis layer, in primary suspension systems.
- At the chassis-to-cabin layer (Figure 1.8), in large vehicles where the driver cabin is separated from the main chassis (e.g. large agricultural tractors, trucks, earth-moving machines, etc.). See Figure 1.8 (left).
- At the cabin-to-seat layer: in large off-road vehicles the driver seat is also frequently equipped with a fully-fledged suspension system, in order to reduce the vibration suffered by the driver during the typically long hours spent in the vehicle (see e.g. ISO2631, 2003). See Figure 1.8 (right).

Many types of vehicle are equipped (or are being equipped) with semi-active suspension; the list is long, multi-faceted, and continuously increasing. Such vehicles range from small vehicles like motorcycles, ATVs, snowmobiles, etc. to large off-road vehicles (agricultural tractors, earth-moving machines, etc.), passing through classical cars, and duty-vehicles such as trucks, ambulances, fire-trucks, etc. (see e.g. Ahmadian and Simon, 2001; Aubouet et al., 2008; Choi et al., 2000; Codeca et al., 2007; Deprez et al., 2005; Fischer and Isermann, 2003; Goodall and Kortum, 2002; Ieluzzi et al., 2006; Spelta et al., 2010).

If we look inside a semi-active damper, today there are three main available technologies, which allow a fast-reacting electronically controlled modification of the damping ratio of a shock absorber (see Figure 1.9):

Figure 1.8: **Examples of chassis-to-cabin (by Same Deutz-Fahr) and cabin-to-seat (by SEARS) semi-active suspension systems.**

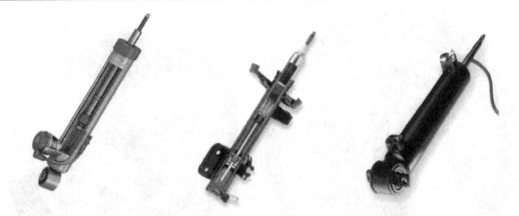

Figure 1.9: Examples of electronically controlled semi-active shock absorbers, using three different technologies. From left to right: solenoid-valve Electrohydraulic damper (Sachs), Magnetorheological damper (Delphi), and Electrorheological damper (Fludicon).

- The (classical) electrohydraulic (EH) technology, based on solenoid valves located inside or outside the main body of the damper; they can change the damping ratio by modifying the size of orifices.
- The magnetorheological (MR) technology, based on fluids which can change their viscosity when exposed to magnetic fields.
- The electrorheological (ER) technology, based on fluids which can change their viscosity when exposed to electric fields.

All these technologies are suitable for vehicle applications. Such technologies today are in strong competition on the basis of many features and parameters, such as: response time, controllability range, stick-slip, fault management, long-term reliability, cost, weight and packaging, maintenance requirements, power-electronics requirements, etc. Each technology has its pros and cons, and none of them provide the best features over all these characteristics.

Looking to the future, three trends of evolving semi-active suspensions can be outlined:

- **Technologies**: The three main technologies available today are still evolving. Another almost ready technology is air-damping with electronically controlled valves. The real "quantum leap", however, will be the replacement of classical fluid-based damping technologies with electric motors, capable of energy recuperation. This technology is particularly attractive since it can integrate semi-active and active capabilities.
- **New vehicle architectures**: The end of the first decade of the new millennium has been characterized by an impressive acceleration of electric car technologies. Pollution, the greenhouse effect, and shortage of fossil fuel have boosted the interest in electricity for

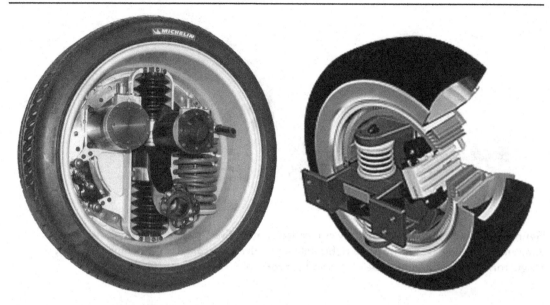

Figure 1.10: Examples of "full-corner" vehicle architectures: Michelin Active Wheel© (left) and Siemens VDO e-Corner© (right).

short-range mobility. Within this mainstream, a completely new vehicle architecture is emerging: the "full-corner" vehicle, where all the main dynamic elements of the vehicle are packed in the wheel: the main electric motor (with energy-recuperation capability), the electro-mechanical "by-wire" brake, the (possibly electronic) suspension, and (possibly) the electro-mechanical "by-wire" steer. Two examples of such "all-in-wheel" devices have been recently presented (see Figure 1.10) by Michelin and Siemens VDO. Interestingly enough, semi-active suspensions will play an even more important role in such architectures, since the comparatively large unsprung mass will worsen the problem of finding a good compromise between comfort and handling.

- **Centralized control strategies**. Last but not least, there is an unstoppable trend also from the control-algorithm point of view. This trend is concisely called "Global Chassis Control" (GCC – see e.g. Gáspár et al., 2007; Poussot-Vassal, 2008), and consists in the integrated and coordinated design of the control strategies of all the vehicle dynamics control subsystems: braking control, traction control, stability control, suspension control and, more recently, kinetic energy management. These subsystems, traditionally designed and implemented as independent (or weakly interleaved) systems, will be increasingly designed in a centralized fashion, in order to fully exploit the potential benefits coming from their interconnection. Again, the capability of designing sophisticated semi-active control algorithms will increase the importance of this trend.

1.4 Book Structure and Contributions

The main objective and contribution of this book is to present, in a condensed and homogeneous form, all the material accumulated in the industrial (patents) and scientific literature on this topic in the last decade. Moreover, most of the book focuses on the design "methods" more than on specific solutions and technologies; semi-active suspension design will be a topic of primary importance in vehicle-dynamics control for many years to come, and – as in other engineering disciplines – the methods last much longer than the specific solutions, which are much more linked to the technologies and requirements of the moment.

The structure of this book follows the classical path of control-system design: first, the actuator (the variable-damping shock absorber) is discussed and modeled; then the vehicle is

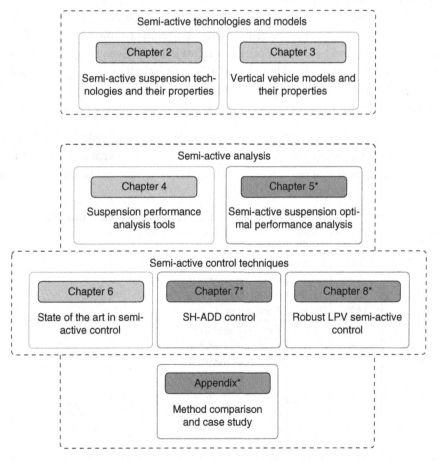

Figure 1.11: Book organization and suggested reader roadmap. Expert readers may start directly with starred (∗) chapters.

mathematically modeled, and finally the control algorithms are designed and discussed. In order to be effectively accessed at different reading levels, the book has been conceptually divided into two parts. In the first part of the book (Chapters 2 to 6) the basics of modeling and semi-active control design are described, whereas in the second part of the book (Chapters 7 to 8 and Appendix) more advanced and research-oriented solutions are proposed and compared, with the help of some case studies.

Consistent with the methodology-oriented flavor of this book, a lot of emphasis has been put in two pivotal chapters: Chapter 4 ("Methodology of analysis for automotive suspensions") and Chapter 5 ("Optimal strategy for semi-active suspensions and benchmark"):

- In Chapter 4 the different techniques and methods for evaluating the performance of a suspension system are discussed in detail, in order to have a common baseline to assess and compare the quality of different design solutions.
- In Chapter 5 an "ideal" semi-active control strategy is developed, by assuming full knowledge (past and future) of the road profile, and using a sophisticated off-line numerical optimization based on model predictive control. Even though this control strategy cannot be implemented in practice, it is conceptually very important since it sets an absolute standard for the best possible filtering performance of semi-active suspensions, and represents a simple and clear benchmark for any "real" algorithm.

Figure 1.11 summarizes the book chapter contents and proposes a roadmap for the reader, depending on a basic knowledge of the semi-active suspension control field. Note that this roadmap may also be used as a basis for a lecture given to an automotive and control engineer

Table 1.1: Automotive parameters set (passive reference model)

Symbol	Value	Unit	Meaning
M	400	kg	Sprung mass
m_{ij}	50	kg	Unsprung masses (i = front, rear and j = left, right)
I_x	250	kg.m^2	Roll inertia
I_y	1400	kg.m^2	Pitch inertia
t	1.4	m	Front and rear axle
l_f	1.4	m	COG-front distance
l_r	1	m	COG-rear distance
r	0.3	m	Nominal wheel radius
h	0.7	m	Chassis COG height
k_f	30,000	N/m	Front suspension linearized stiffness (left, right)
k_r	20,000	N/m	Rear suspension linearized stiffness (left, right)
c_f	1500	N/m/s	Front suspension linearized damping (left, right)
c_r	3000	N/m/s	Rear suspension linearized damping (left, right)
k_t	200,000	N/m	Tire stiffness (front, rear and left, right)
β	50	rad/s	Suspension actuator bandwidth

Table 1.2: Motorcycle parameters set (passive reference model)

Symbol	Value	Unit	Meaning
M	117	kg	Sprung mass
m_i	30	kg	Unsprung masses ($i =$ front, rear)
I_y	1000	kg.m^2	Pitch inertia
l_f	0.75	m	COG-front distance
l_r	0.65	m	COG-rear distance
r	0.3	m	Nominal wheel radius
h	0.5	m	Chassis COG height
k_f	26,000	N/m	Front suspension linearized stiffness
k_r	30,000	N/m	Rear suspension linearized stiffness
c_f	2000	N/m/s	Front suspension linearized damping
c_r	3000	N/m/s	Rear suspension linearized damping
k_t	200,000	N/m	Tire stiffness (front, rear)
β	50	rad/s	Suspension actuator bandwidth

classroom. Finally, it may be used to illustrate some of the modern control methods on an application.

1.5 Model Parameter Sets

In this book, two model parameters sets are considered. The first one represents a classical parameter set for automotive applications (see Table 1.1), while the second one is motorcycle oriented (see Table 1.2). These parameters are very generic and represent the parameters of standard cars and motorcycles. Note that these parameter sets, denoted as '"passive"', will be used throughout the book, to build reference models.

CHAPTER 2

Semi-Active Suspension Technologies and Models

The aim of this chapter is to introduce a description of suspension systems starting from the common passive framework towards the classification of electronically controlled suspensions, with particular attention to semi-active suspensions. In a semi-active suspension system, the variation of damping may be achieved by introducing modulable mechanisms in the shock absorber such as solenoid valves (electrohydraulic dampers), or by the use of fluids which may vary their viscosity if subject to an electric or magnetic field (electrorheological and magnetorheological dampers). A simple but effective description of this kind of actuator is extremely important for control design purposes.

The outline of the chapter is as follows: in Section 2.1 a brief introduction to the suspension system modeling is presented and the common devices for passive suspensions are described in Section 2.2. The classification of electronic suspension systems is conceptually presented in Section 2.3. The electrohydraulic (EH), magnetorheological (MR) and electrorheological (ER) technologies for semi-active suspensions are presented in Section 2.4. The dynamic characterization and modeling of semi-active shock absorbers are dealt with in Section 2.5.

2.1 Introduction to Suspension Modeling

The graphical representation of a suspension system in a vehicle is reported in Figure 2.1, where the so-called quarter-car model is depicted (see e.g. Isermann, 2003). The quarter-car aims at describing the interactions between the suspension system, the tire and the chassis in a single corner of a vehicle.

As is evident from Figure 2.1, the quarter-car representation consists of four simplified elements:

- The suspended mass representing the chassis.
- The unsprung mass that comprises devices such as the wheel mass, the brake, the caliper, etc.
- The tire that is modeled as an elastic element.

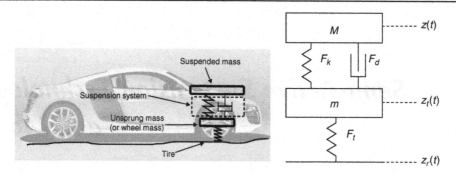

Figure 2.1: Quarter-car representation of a suspension system in a vehicle.

• The suspension system which consists of an elastic element and a dissipative element. The contributions of the elastic and dissipative elements are assumed to be additive.

Definition 2.1 (*Nonlinear quarter-car vertical model*).   *By balancing the forces involved in the quarter-car system (as represented in Figure 2.1), it is possible to write the following set of second-order dynamical equations:*

$$\begin{cases} M\ddot{z}(t) = F_k(t) + F_d(t) - Mg \\ m\ddot{z}_t(t) = -F_k(t) - F_d(t) + F_t(t) - mg \end{cases} \tag{2.1}$$

The symbols used in (2.1) and in Figure 2.1 have the following meaning: z, z_t and z_r are the vertical position of the suspended mass, the vertical position of the unsprung mass, and the road profile, respectively; M and m are the chassis mass and unsprung mass, respectively; F_k, F_d are the forces delivered from the elastic element and the dissipative element of the suspension, respectively; F_t is the force delivered from the elastic element that links the unsprung mass and the ground.

Equations (2.1) represent a fourth-order nonlinear dynamical system. Note that in a passive suspension system, these equations are complemented by the following condition:

Definition 2.2 (*Passivity constraint*).   *In the quarter-car model (2.1), the passivity constraint of the suspension system is given by (gray area in Figure 2.2):*

$$\begin{cases} F_k \cdot x \geq 0 \\ F_d \cdot \dot{x} \geq 0 \\ \quad x = -(z - z_t - L) \end{cases} \tag{2.2}$$

where x is the suspension deflection which is positive if the suspension is compressed, and negative if the suspension is extended. L is the unloaded length of the elastic element in the suspension so that $F_k(t) = 0$, when $x = 0$.

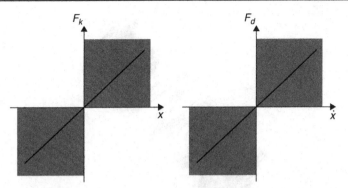

Figure 2.2: Pictorial representation of the suspension "passivity constraint" (gray area). Example of linear characteristics for passive spring (bold line, left) and for passive damper (bold line, right).

From a mechanical perspective, the passivity constraint imposes the elements in the suspension not to introduce energy into the system; the elastic elements may only store the energy transmitted from the road profile, while the damping dissipates it.

Definition 2.3 (*Linear passive suspension*). *The ideal linear damping unit and the ideal linear elastic unit of a suspension are described by the following equations (Figure 2.2):*

$$\begin{cases} F_k = k \cdot x \\ F_d = c \cdot \dot{x} \end{cases} \tag{2.3}$$

where x is the suspension deflection, k is the stiffness of the linear elastic element and c is the damping ratio of the linear dissipative element.

A formal presentation of quarter-car system with extensive analysis is further discussed in the next chapter.

2.2 Passive Suspension Systems

This section presents the devices that are used for the elastic and the dissipative unit of passive suspension. Coil and gas springs are usually adopted as elastic elements in the suspensions for vehicles: the coil springs exploit the elasticity provided by the torsion of coils; gas springs are based on the compressibility of gases. Dissipative devices are implemented by *hydraulic shock absorbers*.

2.2.1 Coil Spring

A coil spring (represented in Figure 2.3) is an elastic element assumed to be subject only to vertical forces and displacements. The force is delivered as a reaction of torsion of the coils, while other secondary effects may be neglected.

Figure 2.3: Example of a steel coil spring.

Definition 2.4 (*Linear coil spring*). *The stiffness of an ideal linear coil spring is given by the following relation (Bosch, 2000):*

$$k = \frac{Gd^4}{32R^3n} \tag{2.4}$$

where G is the rigidity modulus for the material of coils, d is the wire diameter, R is the mean of the internal radius of the coil, n is the number of active coils.

Figure 2.4 represents an example of ideal and realistic characteristics of a coil spring for a standard sedan. By inspecting (2.4) note that, in general, a suspension works "pre-compressed" around a steady state condition, due to the static load of the suspended mass. A coil spring is designed to carry the static load and to let movements occur around the steady state position, both in the compression and in the rebound phase.

2.2.2 Gas Spring

Gas springs are usually implemented in two different suspension frameworks (conceptually depicted in Figure 2.5). The first one is the *pneumatic suspension*, where the spring is implemented by an air chamber surrounding the shock absorber. The second is hydro-pneumatic suspension where an accumulator is linked to the shock absorber and the gas included is compressed by the movements of the fluid.

In order to study the force-deflection characteristic of a gas spring, consider the following assumptions:

* The gas is supposed to be ideal. Note that the air is a fair approximation of an ideal gas.
* During normal use of the suspension, the gas is subject to an adiabatic transformation, so that no thermal energy is exchanged with the ambient. Note that, suspension dynamics are relatively much faster than dynamics of thermal exchanges.

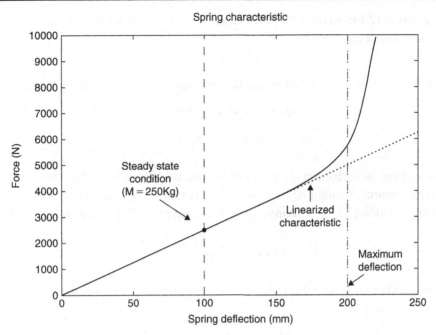

Figure 2.4: Typical deflection-force characteristic (right) of spring with nominal stiffness coefficient $k = 25$ KN and nominal maximum deflection of 200 mm. Steady state computed for a suspended mass of 250 Kg.

- The force delivered by the gas spring is given as a result of the application of a pressure over an active area of the gas chamber. The variations of the active area are supposed to be negligible for any suspension deflection.

Definition 2.5 (*Gas spring characteristic*).   *A gas spring delivers a force F_k proportional to the difference between the internal pressure p and the external ambient pressure p_{atm}, as follows:*

$$F_k = A(p - p_{atm}) \tag{2.5}$$

where A is the active area of the air chamber. The approximated linear stiffness associated to the air spring is given by the following relation:

$$k = \frac{p_0 \gamma A L_0}{(L_0 - x)^{\gamma+1}} \tag{2.6}$$

where p_0 and L_0 are the pressure and the spring length when the suspension is fully expanded (i.e. when it is unloaded), respectively; γ is the adiabatic constant of the gas (for air $\gamma_{air} = 1.4$); x is the suspension deflection ($x \geq 0$), so that $x = 0$ when the suspension is fully compressed, and $x = L_0$ when the suspension is fully expanded.

Proof of equation (2.6): In order to prove equation (2.6), consider the relation of adiabatic transform for an ideal gas:

$$pV^\gamma = p_0 V_0^\gamma$$

where V is the gas volume in the chamber. By differentiating, the following holds:

$$\partial p V^\gamma + p\gamma V^{\gamma-1} \partial V = 0$$

$$\Rightarrow \partial p = -\frac{p\gamma \partial V}{V}$$

(2.7)

Since the variations of deflection are assumed to have no influence on the active area of the spring, the gas volume V might be expressed as $V = (L_0 - x)A$, thus $\partial V = -A\partial x$. By differentiating relation (2.5) and taking into account equation (2.7) the following holds:

$$\partial F = A\partial p = -A\frac{p\gamma \partial V}{V} = \frac{p\gamma A^2}{V}\partial x$$

As the linear stiffness of a spring is defined as $k = \frac{\partial F}{\partial x}$, by algebraic manipulations, it is possible to have proof of (2.6):

$$k = \frac{\partial F}{\partial x} = \frac{p\gamma A^2}{V} = p_0\frac{V_0^\gamma}{V^{\gamma+1}}\gamma A^2$$

$$\Rightarrow k = \frac{p_0\gamma AL_0}{(L_0 - x)^{\gamma+1}}$$

∎

Equation (2.6) reveals that the equivalent linear stiffness associated with a gas spring is not a constant value. It depends on the nominal conditions p_0 and L_0, and on the suspension deflection x. Note that, if the gas spring is fully compressed (i.e. $x \to L_0$), then the delivered

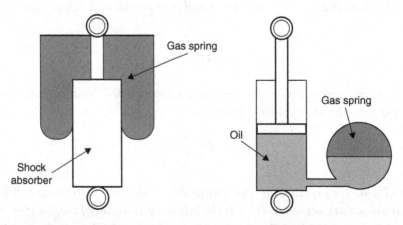

Figure 2.5: Schematic representation of a gas spring implemented with pneumatic spring (left) and with hydro-pneumatic spring (right).

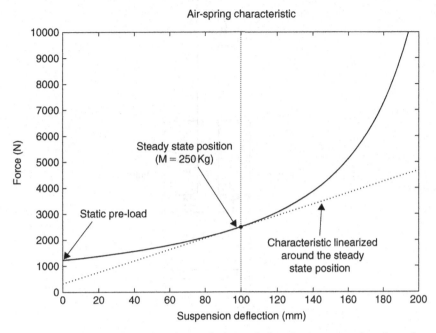

Figure 2.6: Typical deflection-force characteristic of an automotive air spring.

force tends to infinite. Further, it is worth noticing that if the variation deflections are negligible with respect to the nominal length (i.e. $x \ll L_0$) then the stiffness may be considered constant, i.e.

$$k = \frac{p_0 \gamma A L_0}{(L_0 - x)^{\gamma+1}} \approx \frac{p_0 \gamma A}{L_0^{\gamma}}$$

An example of force-deflection characteristic of a gas spring designed for a standard sedan suspension is depicted in Figure 2.6. The parameters used are $p_{atm} = 1$ atm, $p_0 = 80$ atm, and $L_0 = 200$ mm. The static load should be $M = 250$ Kg. From inspecting Figure 2.6, it is evident the regressive behavior of the characteristic and the linear approximation may be considered valid for relatively small deflection variations around the steady state condition.

The characteristic of a gas spring depends on the structural parameters p_0 and L_0. A variation of the characteristic (and of the approximated stiffness) may possibly vary either the nominal pressure or the nominal length. Note that, on this basis, it is thus possible to control the static load (load-leveling systems).

2.2.3 Ideal Damping Element in a Passive Suspension System

The simplified concept of a common passive mono-tube shock absorber is represented in Figure 2.7. A piston is located inside the shock absorber, filled with oil, and it moves according to the suspension deflections. The damping is achieved by the oil that passes through some valves in the piston. The volume of piston rod inside the damper may vary. For volume

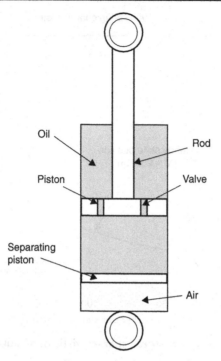

Figure 2.7: Concept of a mono-tube passive shock absorber.

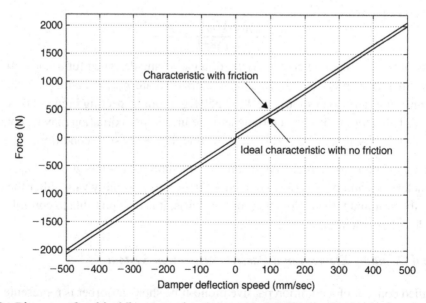

Figure 2.8: Diagram of an ideal linear passive characteristic of hydraulic shock absorber, with and without friction. The damping coefficent is $c = 2000$ Ns/m, the static friction is $F_0 = 70N$.

compensation, an air chamber is included in the damper. Due to both the damping of the piston movements, and due to the elasticity of the air chamber, the force delivered by a shock absorber F_{shock} can be viewed as given by three contributions, ideally assumed as additive:

$$F_{shock} = F_d(x, \dot{x}) + F_{air}(x) + F_0 \textbf{sign}(\dot{x})$$

where F_d is the damping force as a function of suspension deflection x and deflection velocity $\dot{x}$; F_{air} is the force delivered by the gas spring. F_0 represents the internal friction (assumed as constant). Since the contribution of the gas spring is merely an elastic force, this may be logically embedded as a part of the elastic device of the suspension. Thus F_{air} may be neglected in the description of the damping force.

Definition 2.6 (*Ideal linear damper*).   *An ideal linear damper is given by the following force-velocity relation*

$$F_d = c\dot{x} + F_0 \textit{sign}(\dot{x}) \tag{2.8}$$

where c is the damping coefficient of the damper. A graphical representation of a common shock absorber is depicted on Figure 2.8.

Readers interested in further details on the design of passive shock absorbers are referred to Dixon (2007), and Guglielmino and Edge (2004).

2.3 Controllable Suspension Systems: a Classification

Ideally, a control action may be introduced at different levels of the suspension system: at the level of the dissipative unit, by a modulation of the damping force; at the level of the elastic unit, by a modulation of the spring force; at the full level of the suspension, by replacing both the elastic and the damping devices with a force actuator.

This section presents the ideal classification of controllable suspensions, originally introduced by Isermann (2003), from which we adopt the same nomenclature.

The classification of controllable suspension may be carried out according to the energy input and the bandwidth of the actuator. More specifically, three features may be observed: the controllability range, in other words the range of forces that the actuators may deliver; the control bandwidth which is a measure of how fast the actuator action is; the power request that is mainly due to the mix of controllability range and control bandwidth. The resulting classification is concisely shown in Table 2.1, which highlights five families of controllable suspensions:

- **Adaptive suspension**: the control action is represented by a relatively slow modulation of damping, so the control range is limited by the passivity constraint. The adaptive shock absorber is characterized by a bandwidth of a few Hertz. Since no energy is introduced in

Table 2.1: Classification of electronically controlled suspension

System class	Control range (spring)	Control range (damper)	Control bandwidth	Power request	Control variable
Passive			–	–	–
Adaptive			1–5 Hz	10–20 W	*c (damping ratio)*
Semi-active			30–40 Hz	10–20 W	*c (damping ratio)*
Load leveling			0.1–1 Hz	100–200 W	*W (static load)*
Slow active			1–5 Hz	1–5 kW	*F (force)*
Fully active			20–30 Hz	5–10 kW	*F (force)*

the system and the bandwidth is relatively small; the power request is usually limited to a few Watts.

- **Semi-active suspension**: this system features an electronic shock absorber which may vary the damping with a relatively large bandwidth (usually around 30–40 Hz). The deliverable forces follow the passivity constraint of the damper, thus no energy may be introduced into the system. Due to these features, the requested power is relatively low, around tens of Watts.

- **Load-leveling suspensions**: this class of suspensions may be considered the first attempt at active suspensions, since they are capable of introducing energy in the system to change the steady state condition (as a response of a variation of the static load). The control acts on the parameter of the springs (usually an air spring). The bandwidth is usually within 0.1–1 Hz, but the power request is usually some hundreds of Watts.

- **Slow-active suspensions**: in active suspensions, the passivity constraint is fully overcome and energy may be injected into the system. The control input is the suspension force F delivered by an actuator which replaces the passive devices of the suspension. The bandwidth is limited to a few Hertz. Note that the vast controllability range is paid in terms of power request (around some kilo-Watts).
- **Fully-active suspensions**: the difference between slow and fully-active suspensions is in terms of bandwidth. The fully-active actuator is able to react in milliseconds (bandwidth of 20–30 Hz). As in slow-active suspension systems, the control variable is the suspension force F. Interestingly enough, the available bandwidth is the same as semi-active suspensions. However since the controllability range is beyond the passivity constraint, the overall power request is relatively high-demanding, around tens of kilo-Watts.

For control purposes, it is interesting to understand what are the vehicle dynamics that can be managed by the different controllable suspension systems. To this aim, consider that usually a vehicle presents two main vertical dynamics (see next chapter for details). The *body dynamics* are characterized by a bandwidth around 1–5 Hz; the *wheel dynamics* are concentrated around 15–20 Hz. Figure 2.9 provides a representation of the controllable suspension families over the control bandwidth domain, with respect to the power request. With reference to Figure 2.9 some observations may be made:

- The load-leveling suspension may regulate the static load (relatively small bandwidth) but have no influence on the body and wheel vertical dynamics of the vehicle.

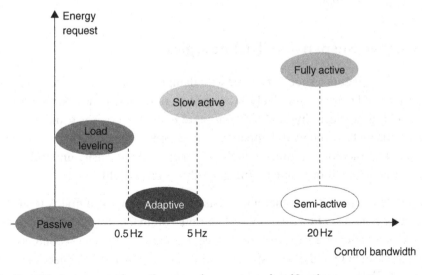

Figure 2.9: Graphic representation of suspension system classification: energy request with respect to the available control bandwidth.

- Adaptive and slow active suspensions are appropriate to control the body dynamics but with no action on the wheel dynamics.
- Semi-active and full-active suspensions are able to control the overall vertical vehicle dynamics. Note that, due to the relatively vast range of deliverable forces, in principle, active suspensions may provide the best performance. However, this is paid for in terms of an extremely large energy requirement and in terms of stability of the system; in fact, by introducing energy into the active systems, unstable behaviors might arise if not appropriately controlled. On the other hand, semi-active suspensions are always stable in a closed-loop system due to the passivity constraint (see next chapter for details).

As a final consideration, in terms of power, requirement control action (controllability range and available bandwidth), intrinsic stability (i.e. safety), semi-active suspensions seem to be the best compromise between achievable performances and costs. In practice, nowadays, the best performances seem to be given by those suspension systems equipped with both a semi-active damping control and a load-leveling.

Remark 2.1 (*Modulation of spring*).   *Although active and damping control solutions have been extensively studied, the control of spring stiffness is a much more subtle and elusive problem. The classical load-leveling or slow-active suspensions based on hydro-pneumatic or pneumatic technologies are subject to spring-stiffness variations, but the stiffness change is more a side-effect than a real control variable. The classification so far focused on available technologies extensively studied and adopted as industrial solutions. Sophisticated and less well-known solutions devoted to the modulation of spring coefficient have been recently presented (Spelta et al., 2009).*

2.4 Semi-Active Suspension Technologies

Semi-active systems feature a shock absorber with damping ratio that may vary with associated bandwidth of 20–30 Hz. Such a relatively fast variation of damping may be achieved by the following technologies: electrohydraulic dampers are hydraulic devices usually equipped with solenoid valve; magnetorheological dampers and electrorheological dampers consist of shock absorbers filled with rheological fluids, which may change the viscosity under the action of an electric (electrorheological) or magnetic (magnetorheological) field.

The description of a semi-active actuator consists of two features. The first is represented by the force-velocity maps as a function of the electronic command; this is the generalization of the force-velocity map of a passive damper (see e.g. Figure 2.8). The second is the dynamic description, represented by the time response of the damper when it is subject to a variation of damping request. To describe the technologies for semi-active systems, this section focuses mainly on the first feature.

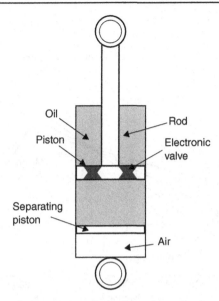

Figure 2.10: Schematic representation of an electrohydraulic shock absorber.

2.4.1 Electrohydraulic Dampers (EH Dampers)

The concept of electrohydraulic damper is depicted in Figure 2.10. Compared to the classical passive element, the electrohydraulic device comprises electronic valves instead of passive valves.

Definition 2.7 (*Ideal characteristics of EH dampers*).   *The ideal characteristic of electrohydraulic dampers is represented by the following relation:*

$$\begin{cases} F_d(\dot{x}, I) = c(I)\dot{x} + F_0 \boldsymbol{sign}(\dot{x}) \\ \quad\quad c(I) = \gamma I + c_0 \\ \quad\quad 0 \leq I \leq I_{max} \end{cases} \tag{2.9}$$

where x is the suspension deflection, I is the electrical command (e.g. a current for solenoid valves) limited between $I = 0$ and $I = I_{max}$. c_0 is the minimum damping achieved when the electronic command is off; γ is the characteristic gain that turns the electronic command into a damping; F_0 is the internal friction of the shock absorber, herein assumed to be constant. A graphical representation of the ideal EH characteristic for vehicles is given in Figure 2.11, where three maps are depicted: minimum damping map, medium damping map, and maximum damping map.

2.4.2 Magnetorheological Dampers (MR Dampers)

Magnetorheological dampers (MR dampers) exploit the physical properties of magnetorheological fluids (MR fluids). MR fluids change their viscosity when subject

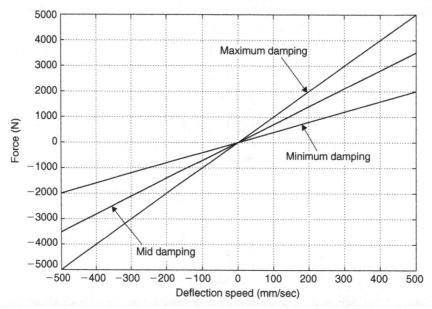

Figure 2.11: Ideal damping characteristics of an electrohydraulic shock absorber (with negligible friction).

to a magnetic field. An MR fluid consists of a mixture of oil (usually a silicon oil) and micro-particles sensitive to the magnetic field (e.g. iron particles). The MR fluid behaves as a liquid when no field is applied. In the case of a magnetic field applied to the fluid, the particles form chains and the fluid becomes very viscous.

In an MR damper (conceptually represented in Figure 2.12) the piston includes coils capable of delivering a magnetic field in the orifices. In these terms, the piston may be viewed as a "magnetorheological valve" and the damping is the result of the friction between the fluid and the orifices.

Definition 2.8 (*Ideal characteristics of MR dampers*). *The ideal and simplified characteristics of MR dampers are represented by the following relation (see e.g. Savaresi et al., 2005a)*

$$\begin{cases} F_d(\dot{x}, I) = c_0\dot{x} + F_{MR}(I, \dot{x}) \\ \quad\quad 0 \leq I \leq I_{max} \end{cases} \tag{2.10}$$

where c_0 is the minimum damping given by the free flowing of the fluid through the piston orifices; x is the suspension deflection; I is the electronic command (usually a current) given to the coils, which must be between $I = 0$ and $I = I_{max}$. F_{MR} is the friction force between the MR fluid and the piston orifices, regarded as a nonlinear function of the current command I and of the deflection speed $\dot{x}$.

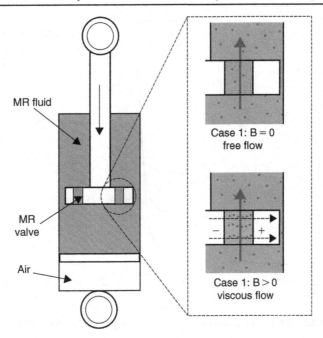

Figure 2.12: Left: schematic representation of a magnetorheological damper behavior: with and without magnetic field (B is the magnetic field).

Model (2.10) is a very concise description of the MR device. Experts in the field know that MR characteristics may be affected by harmful phenomena such as large hysteresis, and variable friction. The literature on MR damper modeling and on their application is extremely vast. For comprehensive presentation see e.g. Guglielmino et al. (2005); Savaresi et al. (2005a); Spencer et al. (1997).

An example of ideal MR damper characteristics is reported in Figure 2.13, for three different levels of electronic command. Notice that the maps are almost flat due to the presence of relatively low minimum damping and relatively high friction phenomena. This almost-flat shape of the force-velocity map is a peculiar feature of the MR damper.

2.4.3 Electrorheological Dampers (ER Dampers)

Similarly to magnetorheological dampers, electrorheological dampers may vary the damping exploiting the physical property of the fluid flowing inside the shock absorber (ER fluid). ER fluids may be viewed as a mixture of oil and micron sized particles which are sensitive to electric field. When no electric field is applied, the fluid is almost free to flow in the damper orifices; when an electric field is applied, the particles react as dipoles and form chains, so that the flowing of the fluid turns from almost free to visco-plastic. An ER damper is conceptually depicted in Figure 2.14. The ER damper can be regarded as a "electric capacitor"; the external

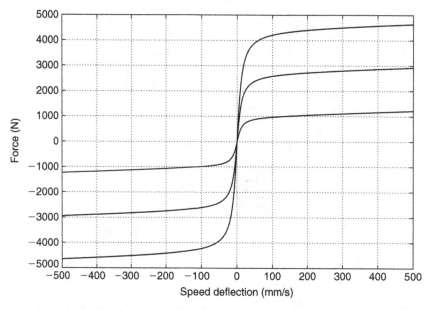

Figure 2.13: **Ideal damping characteristics of a magnetorheological shock absorber.**

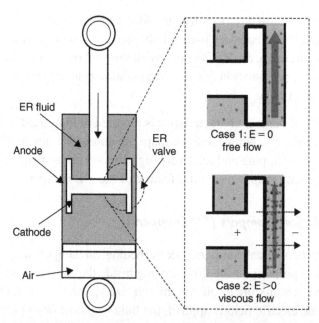

Figure 2.14: **Schematic representation of an electrorheological damper: with and without electric field (E is the electric field).**

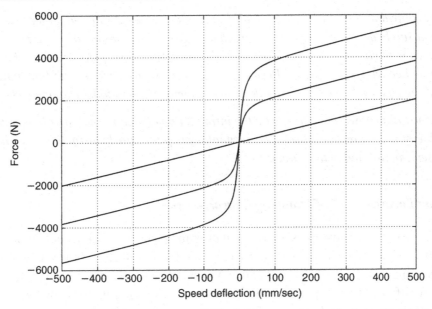

Figure 2.15: Ideal damping characteristics of an electrorheological shock absorber.

body is an anode and the piston is a cathode. The electric field (E) appears in the space between the piston and the body. In order to obtain the necessary forces, the surface of the piston is relatively large. The side-effect is a minimum damping level usually greater than the minimum damping level provided by the MR damper. On the other hand, the friction effect is less harmful and the ER fluid seems to be less aggressive.

Definition 2.9 (*Ideal characteristics of ER dampers*).   *The ideal and simplified characteristics of ER dampers are represented by the following relation (see e.g. Savaresi et al., 2005a)*

$$\begin{cases} F_d(\dot{x}, V) = c_0\dot{x} + F_{ER}(V, \dot{x}) \\ \quad\quad 0 \leq V \leq V_{max} \end{cases} \tag{2.11}$$

where c_0 is the minimum damping given by the free flowing of the fluid through the piston; x is the suspension deflection. V is the electronic command (usually a voltage) given to the piston (cathode) and to the damper body (anode). F_{ER} is the friction force between the ER fluid and the piston surface and is assumed as a nonlinear function of both the deflection speed $\dot{x}$ and the command voltage V. Ideal force-velocity maps for ER dampers are reported in Figure 2.15, for three different levels of electronic command. The shape depicted is a peculiar feature of the ER damper.

Remark 2.2 (*Other upcoming technologies*).   *Looking at the future, the three main technologies available today (EH, MR and ER) are still evolving. Another almost-ready* *technology is air-damping with electronically controlled valves. An air-damping suspension consists in a single device that can be regarded as a gas spring equipped with a piston characterized with variable orifices. Therefore the air-damping suspension mixes the elastic and damping behavior of the gases. Because of this intrinsic dynamical coupling, both the tuning and the electronic control of air-damping suspension seem to be extremely challenging.*

The real innovation however seems to be the introduction of electric motors used as dampers and capable of energy recuperation. This technology is particularly attractive since it can integrate semi-active and active capabilities.

2.4.4 On "linearization" of Damping Characteristics

The characteristics of EH dampers are fairly approximated by an ideal linear damping. For MR and ER dampers, the force-velocity maps are far from linear. However the three technologies are characterized by comparable bandwidth, and also the controllability range is similar: it differs only in terms of shape of the characteristics. Thus, note that with an appropriate electronic control, it is possible to turn the nonlinear maps that characterize ER and MR dampers into linear, similar to those of the EH damper. Under this perspective, the three technologies are comparable. As a consequence, a general concise force-velocity model may be provided, as follows:

$$\begin{cases} F_d(\dot{x}, I) = c(I)\dot{x} + F_0 \text{sign}(\dot{x}) \\ c(I) = \gamma I + c_0 \\ 0 \leq I \leq I_{max} \end{cases} \tag{2.12}$$

where the used symbols are according the notation introduced so far.

2.5 Dynamical Models for Semi-Active Shock Absorber

As described in Figure 2.16, a semi-active shock absorber may be viewed as a system with two inputs and one output. The inputs are the deflection speed $\dot{x}$ and the electronic command for the damping actuation V. The output is the damping force F_d delivered by the suspension. The peculiarity of the shock absorber is that only the deflection speed is a "driving" input: in fact in the case $\dot{x} = 0$, no force is delivered whatever the damping command is.

Looking at the system illustrated in Figure 2.16, it is possible to view the shock absorber as the cascade of two subsystems: one electrical subsystem and one mechanical subsystem.

The electrical subsystem turns the electronic command into a physical signal for damping modulation. For example, in EH dampers the electronic command is a voltage and the physical signal is the current that drives the valve. The mechanical subsystem turns the physical signal

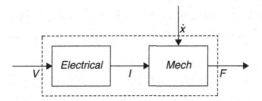

Figure 2.16: Conceptual block diagram of an electronic shock absorber.

electric signal (for damping modulation) into a damping effect. In EH dampers this subsystem is constituted by the moving orifices electronically actuated by the current signal.

2.5.1 Classical Model for Semi-Active Shock Absorber

A full mathematical description of an electronic shock absorber (including the nonlinear behaviors) is not a trivial issue and it has been a topic explored by intensive research activities in the recent past. Models for controllable dampers may be divided into two classes: static models and dynamic models.

- Static model with Coulomb friction:

$$F_d = c(I)\dot{x} - c^{sym}(I)|\dot{x}| + c^{nl}(I)\sqrt{|\dot{x}|}sign(\dot{x}) \tag{2.13}$$

where c, c^{sym} and c^{nl} are parameters depending on the damping command I. This model describes the presence of the static friction as a constant term as function of the deflection speed sign.

- Static model with hysteresis (Shuqui et al., 2006).

$$F_d = A_1(I)\tanh\left[A_3(I)\left(\dot{x} + \frac{V_0(I)}{X_0(I)}x\right)\right] + A_2(I)\left(\dot{x} + \frac{V_0(I)}{X_0(I)}x\right) \tag{2.14}$$

where $\{A_1, A_2, V_0, X_0\}$ are model parameters. Note that since this formulation proposes a damping force function of both deflection and deflection speed, it allows hysteresis phenomena to occur.

- First-order nonlinear dynamical models (Ahmadian and Song, 1999; Koo et al., 2004b). These introduce a dynamical state in the model description. These models are often denoted as Bouc-Wen:

$$\begin{cases} F_d = c_0(I)\dot{x} + k_0(I)(x - x^0) + \gamma(I)z \\ \dot{z} = -\beta(I)|\dot{x}|z|z|^{n-1} - \delta(I)\dot{x}|z|^n + A(I)\dot{x} \end{cases} \tag{2.15}$$

where $\{c_0, k_0, x^0, \gamma, \beta, \delta, A, n\}$ are model parameters (dependent on command input I) and z is the internal state that introduces some dynamics in the model and models hysteresis phenomena.

- Second-order nonlinear dynamical models. The following highly sophisticated model is a generalization of the Bouc-Wen model and it is known in the literature as the Spencer model:

$$\begin{cases} F_d = c(\dot{x})\dot{x} + k_0(I)(x - x^0) + \gamma(I)z + m\ddot{x} + F_0 \\ \dot{z} = -\beta(I)|\dot{x}|z|z|^{n-1} - \delta(I)\dot{x}|z|^n + A(I)\dot{x} \\ c(\dot{x}) = a_1(I)e^{\cdots p(I)} \end{cases} \tag{2.16}$$

where, compared to the parameters already introduced in (2.15), m, F_0, a_1, a_2, p are additional parameters dependent on the command I.

- Nonlinear black-box model (static or dynamic). This kind of description is not based on physical descriptions of the device, but aims at representing the nonlinear input–output dynamical relationship. An example of a black-box model for an electronic shock absorber is the following:

$$F_d(t) = f(x(t), \dot{x}(t), \ddot{x}(t), \ldots, x^{(k)}(t), I(t), \dot{I}(t), \ddot{I}(t), \ldots, I^{(k)}(t)) \tag{2.17}$$

where, in accordance to the notation so far x is the damper deflection and $x^{(i)}$ is the i-th derivative of x; I is the external damping command and $I^{(i)}$ is the i-th derivative of I; k is the regressors order of the inputs x and I. f is a function defined as $f : \Omega \subset \mathbb{R}^{2k+2} \to \mathbb{R}$. f is a universal approximator characterized by parameter vector $\theta \in \mathbb{R}^n$. Several universal approximators have been studied for modeling electronic shock absorbers, such as neural networks, spline, polynomials, etc. (see e.g. Savaresi et al., 2005a).

The above model, with different degrees of complexity and precision, may provide an accurate description of an electronic shock absorber characteristic. From a general perspective, they are addressed as simulation oriented models. Note that the Spencer model (2.16) includes 12 parameters, all dependent on the command I. This makes the above model hard to be tractable from the viewpoint of control design.

2.5.2 Control Oriented Dynamical Model

A control oriented model aims at combining both simplicity and completeness of the description of the semi-active damper dynamics.

Definition 2.10 (*Control oriented dynamical model of semi-active damper*). *A control oriented dynamical model of an ideal semi-active shock absorber is given by the following equations:*

$$\begin{cases} F_d(t, \dot{x}(t)) = c(t)\dot{x}(t) + F_0 \textbf{sign}(\dot{x}) \\ \dot{c}(t) = -\beta_M c(t) + \gamma I(t - \tau) + c_{min} \\ \dot{I}(t) = -\beta_{EI} I(t) + \delta V(t) \\ 0 \leq V(t) \leq V_{max} \end{cases} \tag{2.18}$$

where F_d is the delivered damping force; x is the damper deflection; c is the actual damping ratio; F_0 is the basic friction of the damper; the positive constant β_M represents the bandwidth of the mechanical subsystem; γ is the positive gain characteristic of the damping actuation; I is the damping command; τ is the positive mechanical delay between the command and the actuation of damping; c_{min} is the minimum damping of the shock absorber when the electronic command is off; the positive constant β_{EI} represents the bandwidth of the electric subsystem; δ is the characteristic positive gain of the electrical driver; V is the electronic command (usually a voltage) which is between $V = 0$ (command off) and a maximum command $V = V_{max}$.

This model represents a concise description of shock absorber dynamics. Model (2.18) is a second-order nonlinear dynamical system, whose inputs are the electronic command V and the stroke deflection velocity $\dot{x}$. The output is the damping force F_d. The dynamical states are the damping ratio c and the physical signal for damping actuation I (e.g. the current in the EH valve). The nonlinear behavior is due to the fact that the force output is given by the product between the damping ratio and the exogenous input $\dot{x}$. From Model (2.18) some considerations can be drawn:

- The first and second equations represent the nonlinear mechanical subsystem. The third dynamical equation describes the electrical subsystem modeled as a linear first-order dynamic.
- The model is a generalization of the characteristic maps presented in the previous section. Note that, Model (2.12) is obtained if $\beta_M \to \infty$, $\beta_{EI} \to \infty$, and $\tau = 0$. Namely, if the damping dynamics are infinitely fast.
- It is possible to resemble a standard passive suspension by keeping constant the damping command $V(t) = \bar{V}$.
- The model fulfills the passivity constraint (2.2). In fact due to the first-order dynamics, it is easy to see that if $0 \leq V(t) \leq V_{Max}$ than $c_{min} \leq c(t) \leq c_{Max}$ (where c_{Max} is the maximum damping achieved if $V = V_{Max}$). Thus the product $F_d(t)\dot{x}(t)$ is always positive.

2.5.2.1 Electric Driver of Damping Ratio

In a semi-active shock absorber the electrical subsystem aims at driving the damping in response to an electronic command from the electronic control unit (usually a voltage modulated with PWM). In EH dampers the output is a current (which drives the solenoid valve); in MR dampers the output is a current (which generates the magnetic field). In ER dampers the low voltage input is highly amplified to generate the electric field.

As given by Model (2.18) the electric subsystem is assumed to be a first-order linear system. An electrical representation is so given in Figure 2.17 as a series of a resistance (the electrical load) and an inductance (the reactive element): the voltage V is the electronic command and the current I is the signal that drives the damping. Thus the damping command is dynamically

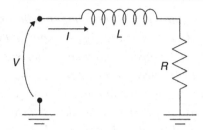

Figure 2.17: Diagram of the electric driver in a semi-active shock absorber.

ruled as follows:

$$L\dot{I}(t) + RI(t) = V(t)$$

$$\dot{I}(t) = -\frac{R}{L}I(t) + \frac{1}{L}V(t) \tag{2.19}$$

Note that from equation (2.18) it is possible to write that $\beta_{EI} = R/L$ and $\delta = 1/L$. By transforming (2.19) according to Laplace, it is possible to obtain the transfer function $G(s)$ from the input V to the output I:

$$I(s) = \frac{1/R}{1 + sL/R}V(s) \tag{2.20}$$

$$G(s) = \frac{I(s)}{V(s)} = \frac{1/R}{1 + sL/R} \tag{2.21}$$

An example of parameters for the electrical driver of semi-active suspension are $L = 30\,\text{mH}$ and $R = 5\,\Omega$ (see e.g. Savaresi et al., 2008). An example of the step response of $G(s)$ is depicted in Figure 2.18.

2.5.2.2 Control of Electric Driver

In order to prevent variations of the electrical parameters (due to drifts) and to speed up the time response, a control loop control is usually implemented in the ECU based on the measure of the current I.

For the sake of simplicity, since the electric driver is modeled as a first-order linear dynamics, a proportional-integral controller is adopted (PI-controller – see for details Årzén, 2003; Åström and Wittenmark, 1997). It is well known that the presence of integral action in the controller eliminates the steady state error, due to, for example, the drifting of electrical parameters.

The transfer function $R(s)$ of ideal PI controller is given as follows:

$$R(s) = K_p + K_I\frac{1}{s} = K_I\frac{K_p/K_I s + 1}{s}$$

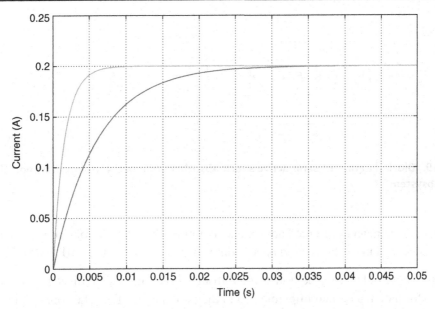

Figure 2.18: Step response of the electric driver: open-loop (bottom line) and closed-loop (top line). Parameters of the driver and the controller are: $L = 30\,mH$; $R = 5\Omega$; desired closed-loop bandwidth $\omega_c = 100 \cdot 2\pi$ (100 Hz); $K_I = 500 \cdot 2\pi$; $K_p = 3 \cdot 2\pi$.

where K_p and K_i are the two tuning knobs of the controller and stand for the proportional and integral gains, respectively. The resulting open-loop transfer function is then given by:

$$L(s) = G(s)R(s) = K_I \frac{K_p/K_I s + 1}{s} \frac{1/R}{L/Rs + 1}$$

If $K_p/K_i = L/R$ a zero-pole cancellation occurs and

$$L(s) = \frac{K_I/R}{s}$$

The resulting closed-loop transfer function $F(s)$, from the reference command I^0 to the actual current I is:

$$F(s) = \frac{L(s)}{1 + L(s)} = \frac{K_I/R}{s + K_I/R}$$

$F(s)$ represents a stable system (one pole with negative real part) with bandwidth $\omega_c = K_I/R$.

In conclusion, the parameter of the PI controller for system (2.21) may be set according to the following relation:

$$\begin{cases} K_p/K_i = L/R \\ K_I = \omega_c R \end{cases} \tag{2.22}$$

Where ω_c is the desired bandwidth of the closed-loop system.

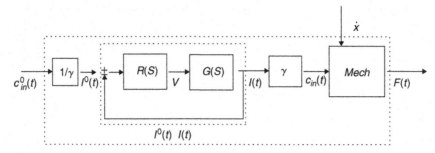

Figure 2.19: Block diagram of semi-active shock absorber equipped with internal control of electric subsystem.

For the driver parameters presented above, and assuming that the control unit runs at a frequency of $\omega_c = 1\,\text{kHz}$, the desired bandwidth is usually chosen at $\omega_c/10 = 100\,\text{Hz}$.

The resulting control law may be transformed from continuous time to discrete time with Tustin transformation and then implemented in the digital controller. The concise performance evaluation of the closed-loop control of current is reported in Figure 2.19, which depicts the step response to a request of 0.2 Amps. Note that the 90%-response time is reduced from 20 ms to 5 ms.

Remark 2.3 (*On the limits of the controller*). *In principle the requested bandwidth ω_c may be as large as possible (limited to the Nyquist frequency due to digitalization). The drawback of a large bandwidth is that the control action requested might be relatively high. This may be in contrast to the limits of the electric actuation (e.g. 12 V in vehicles such as cars and motorcycles). The presence of this saturation (i.e. nonlinearities) coupled with the integral action of the controller may introduce harmful overshoots and delay in the closed-loop time response. The problem may be partially solved with an anti-windup control framework (see e.g. Årzén, 2003; Åström and Wittenmark, 1997; Roos, 2007).*

2.5.3 Simplified First-Order Model for Semi-Active Shock Absorber

In some semi-active shock absorbers the electrical subsystems have dynamics much faster than the mechanical ones (i.e. $\beta_{EI} \gg \beta_M$). This is true for common dampers with a bandwidth of $\beta_M = 20 - 30 \cdot 2\pi$ and a closed-loop servo loop to manage the current command. If the dynamic of the electric driver is neglected (with respect to the mechanical response), a simplified model may be adopted, as represented in Definition 2.11:

Definition 2.11 (*First-order dynamical model of semi-active damper*). *Starting from (2.18) with some algebraic manipulation, by neglecting the electric dynamics, and by omitting the effect of friction, it is possible to define a first-order simplified dynamic model for ideal* *electronic-shock absorbers:*

$$\begin{cases} F_d(t, \dot{x}(t)) = c(t)\dot{x}(t) \\ \quad\quad \dot{c}(t) = -\beta_M c(t) + \beta_M c_{in}(t - \tau) \end{cases} \tag{2.23}$$

where $c_{in}(t) = \gamma I$ and $c_{min} \leq c_{in} \leq c_{max}$. The symbols used are the same as in (2.18).

This very concise model is the basis of vehicle suspension modeling and suspension control design presented in the following chapters.

2.6 Conclusions

In this chapter a general presentation of suspension systems (passive and controllable) has been provided. In general a controllable suspension system may be classified according to the bandwidth of actuation and the energy that the suspension might introduce into the system. According to these criteria, five classes can be defined: fully-active, slow-active, load-leveling, semi-active and adaptive-damping. Among them, semi-active suspension may be considered as the best compromise between achievable performance and cost. Semi-active suspensions are implemented by the use of controllable shock absorbers. The technologies available are based on devices with variable orifices (electrohydraulic dampers) or on devices with fluids capable of varying their viscosity as a function of electric or magnetic field (electrorheological and magnetorheological dampers). These technologies are comparable in terms of bandwidth and controllability range, except for the peculiar shape of the characteristics of each kind of shock absorber. A semi-active damper may be viewed as a system with two inputs (suspension deflection and damping request) and one output (damping force). For control design, an effective model has been presented as a linear model of the damping force with a first-order actuation of the damping request (with delay).

CHAPTER 3

Suspension Oriented Vehicle Models

The present chapter is devoted to the presentation of the main models to analyze the suspension systems, together with their structural limitations and properties. The presented models are general and focus on the vertical vehicle dynamics, directly influenced by the suspension system. Longitudinal, lateral and yaw dynamics are not considered in the book (except for the introduction of the global-chassis-control), as the suspension systems have no or only secondary effects on them. Interested readers are referred to specific works by Gillespie (1992); Kiencke and Nielsen (2000); Milliken and Milliken (1995) books; Poussot-Vassal (2008); Ramirez-Mendoza (1997); Sammier (2001) PhD thesis. This chapter aims at being didactic and self-contained. Models are introduced in a control-oriented perspective with increasing complexity; the definitions are given from a general point of view, then simplified to fit the study objectives and to be understood by unfamiliar readers.

The chapter is organized as follows: Section 3.1 presents the well-known passive vertical quarter-car model, with an analysis of some notable properties and limitations supported with numerical discussions. Passive vertical half-car models, including pitch additional dynamics, is introduced in Section 3.2 with analysis and additional properties. Then, Section 3.3 provides the natural extension of the previous models, defining the full vertical car model. In Section 3.4, an extended half-vehicle model is presented, including the longitudinal dynamics as well, in order to illustrate the existing link between suspension control and longitudinal dynamics (this model may be viewed as a full vertical longitudinal motorcycle model). The semi-active extension of the quarter-car model is given in Section 3.5, introducing the models used throughout the book for controller synthesis.

3.1 Passive Vertical Quarter-Car Model

As already introduced in Definition 2.1 of the previous chapter, when suspension modeling and control are considered, the well-known vertical quarter-car model is often used. This model allows us to study the vertical behavior of a vehicle according to the suspension type (passive or controlled). In this book, this model is largely used for control design and for performance analysis. In Figure 3.1, the passive (both general and simplified forms) quarter-car model is shown.

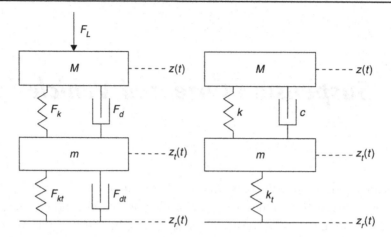

Figure 3.1: Passive quarter-car model, general form (left) and simplified form (right).

In the following, according to the notation of Figure 3.1 (see also Table 1.2), the mathematical models of the vertical quarter-car are derived and analyzed.

3.1.1 Nonlinear Passive Model

Let us first define in a very general way the passive quarter-car model as follows:

$$\begin{cases} M\ddot{z}(t) = F_k(t) + F_d(t) + F_L(t) - Mg \\ m\ddot{z}_t(t) = -F_k(t) - F_d(t) + F_{kt}(t) + F_{dt}(t) - mg \end{cases} \tag{3.1}$$

where $F_k(t)$ and $F_d(t)$ are functions describing the suspension spring and damper vertical forces, respectively. $F_{kt}(t)$ and $F_{dt}(t)$ are functions describing the tire stiffness and damping vertical forces respectively. The gravitational constant is denoted by g. M and m are the chassis and wheel masses, respectively. $z(t)$ and $z_t(t)$ are the vertical chassis and wheel bounce absolute displacements. Finally, $z_r(t)$ and $F_L(t)$ are respectively the road vertical disturbance and vertical load disturbances.

In this model, the unsprung mass m corresponds to the set of elements that compose the wheel, the suspension system, and multiple links from the chassis to the road. Without loss of generality, we will refer to as the wheel since $z_t(t)$ denotes the center of the wheel position.

The model defined in Definition 2.1 is directly obtained by the application of the fundamental dynamical law. The reader should bear in mind that in reality $F_k(t)$, $F_d(t)$, $F_{kt}(t)$ and $F_{dt}(t)$ forces should be described as dynamical nonlinear multivariable functions. But since this book is focused on semi-active suspension control design and analysis only, the following assumptions will be considered:

- The vertical load disturbance is not considered as an input of the system. It results in:

$$F_L(t) = 0 \tag{3.2}$$

Indeed, $F_L(t)$ is used to model load transfer effects occurring when steering and braking maneuvers are performed. For the sake of clarity, in this book, these effects are not considered since they are linked to other dynamics not specifically covered here. Note that the suspension system concerns mainly the road unevenness filtering, hence, our analysis will focus on the suspension filtering abilities with respect to this disturbance (see more details in Chapter 4).

- Linear description of the tire spring force, i.e.:

$$F_{kt}(t) = -k_t(z_t(t) - z_r(t) - R_t) \tag{3.3}$$

where $k_t \in \mathbb{R}^+$ and $R_t \in \mathbb{R}^+$ are the linearized stiffness and nominal length (or radius) of the tire spring respectively.

- Negligible tire damping factor, i.e.:

$$F_{dt}(t) = 0 \tag{3.4}$$

Indeed, in practice, the damping coefficient of a tire is much smaller than stiffness and can be ignored. Further this parameter is not of major interest for suspension analysis and control. Additionally, as far as the tire is concerned, it is worth noting that the main challenge lies in the longitudinal and lateral force descriptions, while the vertical one is of poor interest in the literature (see e.g. Burckhardt, 1993; Corno, 2009; Kiencke and Nielsen, 2000; Tanelli, 2007).

- Linear description of the suspension spring force, i.e.:

$$F_k(t) = -k(z(t) - z_t(t) - L) \tag{3.5}$$

where $k \in \mathbb{R}^+$ and $L \in \mathbb{R}^+$ are the linearized stiffness and nominal length of the suspension spring respectively. Note that semi-active suspension aims at normally operating around linear points. Indeed, in this book framework, nonlinearities should not be reached, and even if they are, appropriate mechanisms (often mechanical) should be used to handle these particular cases. The Reader should especially bear in mind the fact that the main semi-active control challenge lies in the dissipative constraint of the damper control (see Chapters 1 and 2) and not in the spring nonlinearities description (see e.g. Zin, 2005).

- Linear description of the suspension spring force, i.e.:

$$F_d(t) = -c(\dot{z}(t) - \dot{z}_t(t)) \tag{3.6}$$

where $c \in \mathbb{R}^+$.

These model simplifications are very convenient for controller synthesis and numerical simulation efficiency. Moreover, as explained, they are not so restrictive for the application purpose. Then, according to these assumptions, the quarter car model is simplified and may be rewritten as (3.7), fitting with the simplified quarter-car model given in Figure 3.1(right) which will be considered from now on.

$$\begin{cases} M\ddot{z}(t) = -k(z(t) - z_t(t) - L) - c(\dot{z}(t) - \dot{z}_t(t)) - Mg \\ m\ddot{z}_t(t) = k(z(t) - z_t(t) - L) + c(\dot{z}(t) - \dot{z}_t(t)) - k_t(z_t(t) - z_r(t) - R_t) - mg \end{cases} \tag{3.7}$$

Model (3.7) is then valid under a more restrictive domain. More specifically, this model can be considered as valid if and only if:

- Deflections of the suspension are small around the nominal load compression, i.e. road disturbances are the only disturbances acting on the system and are small enough not to enter the suspension limitations and nonlinearities: $|z(t) - z_t(t) - L| \leq L_{max}$ (where L_{max} is the deflection limit of the suspension).
- The wheel is always linked to the ground (the vehicle does not jump the road); indeed, when wheel is no longer in contact with the road, the vehicle is not controllable.

3.1.2 Equilibrium Points

From the last passive quarter-car simplified model (3.7), the system equilibrium point is simply derived as follows:

$$\begin{cases} -k(z^{eq} - z_t^{eq} - L) - Mg = 0 \\ k(z^{eq} - z_t^{eq} - L) - k_t(z_t^{eq} - z_r^{eq} - R_t) - mg = 0 \end{cases} \tag{3.8}$$

Consequently, the solution of this system simply is,

$$\begin{bmatrix} z^{eq} \\ z_t^{eq} \end{bmatrix} = \begin{bmatrix} -k & k \\ k & -k-k_t \end{bmatrix}^{-1} \begin{bmatrix} Mg - kL \\ mg + kL - k_t R_t - k_t z_r^{eq} \end{bmatrix} \tag{3.9}$$

Then, by choosing $z_r^{eq} = 0$, the equilibrium point may be rewritten as:

$$\begin{bmatrix} z^{eq} \\ z_t^{eq} \end{bmatrix} = -\frac{1}{kk_t} \begin{bmatrix} k+k_t & k \\ k & k \end{bmatrix} \begin{bmatrix} Mg - kL \\ mg + kL - k_t R_t \end{bmatrix} \tag{3.10}$$

which results in:

$$\begin{bmatrix} z^{eq} \\ z_t^{eq} \end{bmatrix} = \begin{bmatrix} L - \dfrac{Mg}{k} + R_t - \dfrac{(M+m)g}{k_t} \\ R_t - \dfrac{(M+m)g}{k_t} \end{bmatrix} \tag{3.11}$$

This equilibrium point will then be used to simplify again the system model.

3.1.3 LTI Passive Models

According to the equilibrium point, the LTI quarter-car model can be derived. In the following definition, for the sake of readability, the notation has not been changed. Now, $z(t)$ (resp. $z_t(t)$) denotes the variations of the vertical heights of the M (resp. m) mass around the equilibrium point $[z^{eq}; z_t^{eq}]$. Moreover, the time dependency notation (t) will also be omitted for brevity.

Definition 3.1 (Dynamical-passive LTI vertical quarter-car model). *The simplified passive LTI vertical quarter-car model, as depicted on Figure 3.1 (right), is given by the following dynamical equations:*

$$\begin{cases} M\ddot{z} = -k(z - z_t) - c(\dot{z} - \dot{z}_t) \\ m\ddot{z}_t = k(z - z_t) + c(\dot{z} - \dot{z}_t) - k_t(z_t - z_r) \end{cases} \tag{3.12}$$

where $k \in \mathbb{R}^+$ and $k_t \in \mathbb{R}^+$ are the linearized suspension and tire stiffness coefficients, respectively. $c \in \mathbb{R}^+$ is the linearized damping coefficient. M and m are the chassis and wheel masses, respectively. z and z_t are now the vertical chassis and wheel bounce relative displacements. Finally, z_r is the road vertical disturbance.

State Space Representation   Definition 3.1, describes the LTI passive quarter-vehicle model. Then, an associated LTI state space representation can be derived as follows:

$$\begin{bmatrix} \ddot{z} \\ \dot{z} \\ \ddot{z}_t \\ \dot{z}_t \end{bmatrix} = \begin{bmatrix} \dfrac{-c}{M} & \dfrac{-k}{M} & \dfrac{c}{M} & \dfrac{k}{M} \\ 1 & 0 & 0 & 0 \\ \dfrac{c}{m} & \dfrac{k}{m} & \dfrac{-c}{m} & \dfrac{-k-k_t}{m} \\ 0 & 0 & 1 & 0 \end{bmatrix} \begin{bmatrix} \dot{z} \\ z \\ \dot{z}_t \\ z_t \end{bmatrix} + \begin{bmatrix} 0 \\ 0 \\ \dfrac{k_t}{m} \\ 0 \end{bmatrix} z_r \tag{3.13}$$

Some Useful Transfer Functions   Since they will be largely used throughout this book, from the state space representation given in (3.13), the following transfer functions are described (where s stands for the Laplace variable):

- $F_z(s)$ and $F_{\ddot{z}}(s)$, the transfer functions from the road vertical disturbance $Z_r(s)$ to the chassis displacement $Z(s)$ and acceleration $s^2 Z(s)$ are defined as respectively:

$$F_z(s) = \frac{(ck_t)s + k_t k}{(Mm)s^4 + (cm + cM)s^3 + (Mk + mk + Mk_t)s^2 + (ck_t)s + k_t k} \tag{3.14}$$

$$F_{\ddot{z}}(s) = s^2 F_z(s)$$

These transfer functions are usually related to comfort specifications (see Chapter 4). Then it is interesting to note that:

- $F_z(s)$ has a unitary static gain and tends to zero with a slope of $-60\,\text{dB/dec}$ as $s = j\omega$ increases (the chassis follows the road movements).

- $F_{\ddot{z}}(s)$ has null static gain and tends to zero with a slope of $-20\,\text{dB/dec}$ as $s = j\omega$ increases.
- $F_{z_t}(s)$, $F_{z-z_t}(s)$ and $F_{z_t-z_r}(s)$, the transfers from the road vertical disturbance $Z_r(s)$ to the wheel displacement $Z_t(s)$, the suspension deflection $Z_{z_{def}} = Z(s) - Z_t(s)$ and the tire deflection $Z_{z_{def_t}} = Z_t(s) - Z_r(s)$ are defined as respectively:

$$
\begin{aligned}
F_{z_t}(s) &= \frac{(Mk_t)s^2 + (ck_t)s + k_t k}{(Mm)s^4 + (cm + cM)s^3 + (Mk + mk + Mk_t)s^2 + (ck_t)s + k_t k} \\
F_{z-z_t}(s) &= F_z(s) - F_{z_t}(s) := F_{z_{def}}(s) \\
F_{z_t-z_r}(s) &= F_{z_t}(s) - 1 := F_{z_{def_t}}(s)
\end{aligned}
\tag{3.15}
$$

These transfers are usually related to road-holding and suspension limitation specifications (see Chapter 4). Then it is interesting to note that:

- $F_{z_t}(s)$ has a unitary static gain and tends to zero with a slope of $-40\,\text{dB/dec}$ as $s = j\omega$ increases.
- $F_{z_{def}}(s)$ has null static gain and tends to zero with a slope of $-40\,\text{dB/dec}$ as $s = j\omega$ increases.
- $F_{z_{def_t}}(s)$ has null static gain and tends to unitary gain as $s = j\omega$ increases.

From these representations, it is easy to show that when $c = 0$ (i.e. the suspension system is only composed by a spring, without any damping system), the denominator of the previous rational transfer functions presented becomes:

$$
D(s) = Mms^4 + (Mk + mk + Mk_t)s^2 + k_t k
\tag{3.16}
$$

where $Mm > 0$, $Mk + mk + Mk_t > 0$ and $k_t k > 0$, rendering the system purely oscillatory and presenting two resonance peaks; one for each mass (see also numerical analysis of Section 3.1.5).

3.1.4 Quarter-Car Model Invariance Properties

Here, let us introduce some of the remarkable properties of the quarter-car model, as described by Definition 3.1. Among them, the system exhibits some invariant points (or invariant behaviors) with respect to the damping coefficient. They are referred to as invariant since they characterize specific points in the frequency domain that cannot be modified with a passive suspension. The interested reader may study Gillespie (1992); Sammier (2001).

Before introducing the invariant points of interest for the passive quarter-car model, let us first define an invariant point as follows:

Definition 3.2 (*Invariant points*). *Let consider a transfer function given by $F(j\omega, c)$, where $\omega \in \Omega \subseteq \mathbb{R}^+$ and where $c \in \mathbb{R}^+$ is a coefficient entering in the transfer function description.*

Then, the transfer function $F(j\omega, c)$ has an invariant point (or invariant behavior) with respect to the c parameter, iff.

$$\left\{\exists \eta \in \mathbb{R}^+ \text{ and } \omega_0 \in \Omega \text{ such that } \forall c \in \mathbb{R}^+, \ |F(j\omega, c)|_{\omega=\omega_0} = \eta\right\} \tag{3.17}$$

In the following properties (3.1, 3.2 and 3.3), the invariant points of the transfer functions from z_r to z, z_{def_t} and z_{def} are given. Note that these invariant points are of particular interest for the semi-active application.

Property 3.1 (*Quarter-car invariant points for $F_z(j\omega)$*). *The transfer $F_z(j\omega)$ of the quarter-car model, as given in Definition 3.1, has four invariant points in $\omega \in \mathbb{R}^+$, defined as follows:*

$$\begin{cases} \omega_1 = 0 \\ \omega_2 = \dfrac{1}{mM\sqrt{2}}\sqrt{mM(Mk_t + 2Mk + 2mk - \alpha)} \\ \omega_3 = \sqrt{\dfrac{k_t}{m}} \\ \omega_4 = \dfrac{1}{mM\sqrt{2}}\sqrt{mM(Mk_t + 2Mk + 2mk + \alpha)} \end{cases} \tag{3.18}$$

where,

$$\alpha = \sqrt{4m^2k^2 - 4mkMk_t + 8Mk^2m + M^2k_t^2 + 4M^2kk_t + 4M^2k^2} \tag{3.19}$$

Proof of Property 3.1: A sketch of the proof is given here, leading to the four invariant points given in Property 3.1. Let us consider the following transfer function, where $s = j\omega$ is the Laplace variable:

$$F_z(s) = \frac{Z(s)}{Z_r(s)} = \frac{(ck_t)s + k_t k}{(Mm)s^4 + (cm + cM)s^3 + (Mk + mk + Mk_t)s^2 + (ck_t)s + k_t k} \tag{3.20}$$

then, as defined in Definition 3.2, one aims at finding $\eta \in \mathbb{R}^+$ and $\omega \in \mathbb{R}^+$ such that,

$$\frac{|(ck_t)j\omega + k_t k|}{|Mm\omega^4 - (cm + cM)j\omega^3 - (Mk + mk + Mk_t)\omega^2 + (ck_t)j\omega + k_t k|} = \eta$$

$$\Leftrightarrow \frac{(ck_t\omega)^2 + (k_t k)^2}{(Mm\omega^4 - (Mk + mk + Mk_t)\omega^2 + k_t k)^2 + ((ck_t)\omega - (cm + cM)\omega^3)^2} = \eta \tag{3.21}$$

$$\Leftrightarrow \quad (ck_t\omega)^2 + (k_t k)^2 - \eta\big((Mm\omega^4 - (Mk + mk + Mk_t)\omega^2 + k_t k)^2$$
$$+ ((ck_t)\omega - (cm + cM)\omega^3)^2\big) = 0$$

$$\Leftrightarrow \quad \big((k_t k)^2 - \eta(Mm\omega^4 - (Mk + mk + Mk_t)\omega^2 + k_t k)^2\big)$$
$$+ \big((k_t\omega)^2 + \eta(k_t\omega - (m + M)\omega^3)^2\big)c^2 = 0$$

Then, to solve such an equation while guaranteeing the c parameter independency, the following system has to be fulfilled:

$$\begin{cases} (k_t k)^2 - \eta(Mm\omega^4 - (Mk + mk + Mk_t)\omega^2 + k_t k)^2 = 0 \\ \qquad (k_t \omega)^2 + \eta(k_t \omega - (m + M)\omega^3)^2 = 0 \end{cases} \qquad (3.22)$$

which is equivalent to,

$$\left[\omega k_t(\omega^4 Mm - \omega^2(Mk + mk + Mk_t) + kk_t)\right]^2 - \left[kk_t(\omega k_t - \omega^3(m + M))\right]^2 = 0 \qquad (3.23)$$

Then, by solving this last equation (3.23), with $\omega \geq 0$ leads to the solutions given in Property 3.1. ∎

Property 3.2 (*Quarter-car invariant points for $F_{z_t - z_r}(j\omega)$*). *The transfer $F_{z_t - z_r}(j\omega)$ of the quarter-car model, as given in Definition 3.1, has three invariant points in $\omega \in \mathbb{R}^+$, defined as follows:*

$$\begin{cases} \omega_1 = 0 \\ \omega_5 = \dfrac{1}{2\sqrt{mM(M+m)}}\sqrt{4Mmk + 2m^2k + 2mMk_t + 2M^2k + k_t M^2 - \sqrt{\alpha}} \\ \omega_6 = \dfrac{1}{2\sqrt{mM(M+m)}}\sqrt{4Mmk + 2m^2k + 2mMk_t + 2M^2k + k_t M^2 + \sqrt{\alpha}} \end{cases} \qquad (3.24)$$

where,

$$\begin{aligned} \alpha = {} & 4m^4k^2 + 4M^4k^2 + k_t^2 M^4 - 12M^2m^2kk_t - 8m^3kMk_t + 24M^2m^2k^2 + 16Mm^3k^2 \\ & + 16M^3mk^2 + 4m^2 M^2k_t^2 + 4mM^3k_t^2 + 4M^4kk_t \end{aligned} \qquad (3.25)$$

Proof of Property 3.2: The proof is similar to the previous one related to Property 3.1. Here the equation to be solved is:

$$\begin{aligned} & \left[\omega^3(M + m)(\omega^4 Mm - \omega^2(Mk + mk + Mk_t) + kk_t)\right]^2 \\ & -\left[(\omega k_t - \omega^3(M + m))(-\omega^4 Mm + (Mk + mk)\omega^2)\right]^2 = 0 \end{aligned} \qquad (3.26)$$

∎

Property 3.3 (*Quarter-car invariant points for $F_{z - z_t}(j\omega)$*). *The transfer $F_{z_t - z_r}(j\omega)$ of the quarter-car model, as given in Definition 3.1, has two invariant points in $\omega \in \mathbb{R}^+$, defined as follows:*

$$\begin{cases} \omega_1 = 0 \\ \omega_7 = \sqrt{\dfrac{k_t}{M + m}} \end{cases} \qquad (3.27)$$

Proof of Property 3.3: Similarly to proof related to Property 3.1, the invariant points can be derived by solving a complex polynomial, but it can also be proved by noting that:

$$
\begin{aligned}
& k_t(z_r - z_t) + k_t(z_t - z_r) = 0 \\
\Leftrightarrow \quad & k_t(z_r - z_t) + m\ddot{z} - m\ddot{z} + k_t z - k_t z + k_t z_t - k_t z_r = 0 \\
\Leftrightarrow \quad & M\ddot{z} + m_t\ddot{z}_t + m\ddot{z} - m\ddot{z} + k_t z - k_t z + k_t z_t = k_t z_r \\
\Leftrightarrow \quad & ((M+m)(j\omega)^2 + k_t)Z(j\omega) + (m(j\omega)^2 + k_t)(Z(j\omega) - Z_t(j\omega)) = k_t Z_r(j\omega)
\end{aligned}
\tag{3.28}
$$

Then, by noting $\omega = \omega_7 = \sqrt{\frac{k_t}{M+m}}$, then,

$$
\left. \frac{Z_{def}(j\omega)}{Z_r(j\omega)} \right|_{\omega_7} = \left. \frac{Z(j\omega) - Z_t(j\omega)}{Z_r(j\omega)} \right|_{\omega_7} = \frac{M+m}{M}
\tag{3.29}
$$

∎

Remark 3.1 (*Specific invariant points*). *Additionally, the reader should note that the invariant points in $\omega_1 = 0$, $\omega_3 = \sqrt{\frac{k_t}{m}}$ and $\omega_7 = \sqrt{\frac{k_t}{M+m}}$ are also independent from the k parameter, which practically means that whatever the designed suspension (either passive, semi-active or even active), this point is not modified, i.e. these behaviors are also fully independent from both the spring and damping suspension forces (Moreau, 1995). As a matter of fact, these points are even more particular. These properties have to be kept in mind during the suspension control design step.*

The invariant points given here concern the transfer functions from the road to the chassis, to the tire deflection and to the suspension deflection, but other transfers have notable invariant behaviors which are not described here for sake of readability.

3.1.5 Numerical Discussion and Analysis

The present numerical discussion consists in analyzing the previous model properties on a numerical example. The model used in the following simulations is the simple LTI passive one, as given in Definition 3.1 with the motorcycle parameters given in Table 1.2 and with $c \in [c_{min}, c_{max}] = [700, 3000]$ (i.e. damping between minimal and maximal values). Since the main objective is to analyze semi-active suspensions, i.e. suspensions with controlled damping coefficients, the analysis carried in this section is centered on the behavior of the passive system for varying damping values.

Figure 3.2 illustrates the pole location of the passive LTI quarter-car model for different damping values (i.e. $c \in [c_{min}, c_{max}]$). First, according to this figure, it is notable that whatever the values of $c > 0$, the quarter-car system remains stable (i.e. all eigenvalues in the right half continuous plain). Then, by increasing the c coefficient, the poles moduli increase and are more damped. More specifically,

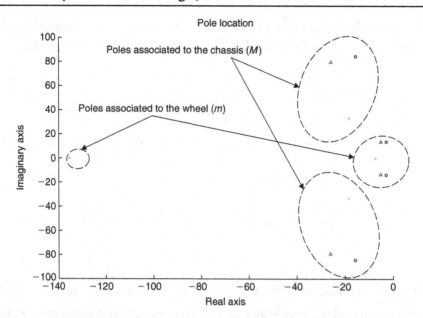

Figure 3.2: Eigenvalues of the passive quarter-car model for varying damping values. Low damping (rounds), medium damping (triangles) and high damping (dots).

- When $c = 0$, the poles are located on the imaginary axis, leading to an oscillating behavior.
- When $c \rightarrow \infty$, the poles tend to be located on the real axis.

On Figures 3.3 and 3.4, the frequency responses of $F_z(s)$, $F_{z_{def_t}}(s)$ and $F_{z_{def}}(s)$ for varying c damping values and k stiffness values are plotted.

From these figures, it is interesting to note the following points:

- First, on Figure 3.3, as expected and accordingly to the pole location analysis provided with Figure 3.2, reducing the damping factor increases the resonance peaks, leading to an oscillatory behavior.
- Then, as explained in Properties 3.1, 3.2 and 3.3, and illustrated on Figure 3.3, the frequency responses of $F_z(s)$, $F_{z_{def_t}}(s)$ and $F_{z_{def}}(s)$ present notable invariant points with respect to the damping value (note that invariant point in $f_1 = 0$ is not visible on the frequency response diagrams, but still exists). More specifically, on this numerical application, the invariant points are given as follows (in frequency domain, $f = \omega/2\pi$):

$$\begin{cases} f_1 = 0\,Hz \\ f_2 \approx 3\,Hz \\ f_3 \approx 13\,Hz \\ f_4 \approx 14.7\,Hz \\ f_5 \approx 2.8\,Hz \\ f_6 \approx 11\,Hz \\ f_7 \approx 5.9\,Hz \end{cases} \tag{3.30}$$

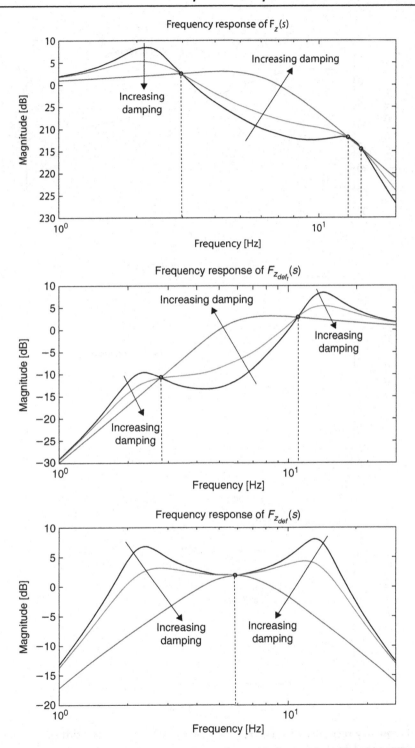

Figure 3.3: Frequency response of $F_z(s)$, $F_{z_{def_t}}(s)$ and $F_{z_{def}}(s)$ for varying damping value c. Invariant points are represented by the dots.

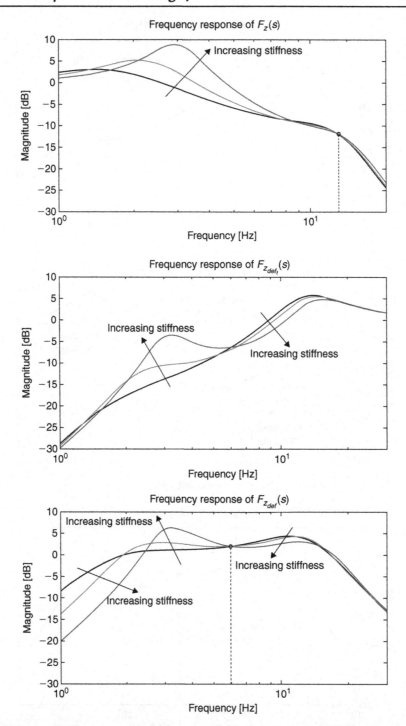

Figure 3.4: Frequency response of $F_z(s)$, $F_{z_{def_t}}(s)$ and $F_{z_{def}}(s)$ for varying stiffness value k. Invariant points are represented by the dots.

Note that these particular points, on transfer $F_z(s)$, will be the basis of the Mixed SH-ADD control design presented later in Chapter 7 and will also be used for $\mathcal{H}_\infty$- "LPV semi-active" control design in Chapter 8.

- On Figure 3.3, the "invariant points" are shown by dots. One must bear in mind that these invariant points are invariant in the linear framework, i.e. if a nonlinear control law is applied most of them would be modified (see e.g. Chapter 5).
- As previously explained, some of these invariant points are invariant with respect to both the damping and the stiffness values; this is the case of $\{f_1, f_3, f_7\}$. Indeed on Figure 3.4, where analysis is shown for varying stiffness k values, these invariant points are also plotted.
- Finally, for low damping values, it is interesting to underline the double resonance which represents the oscillation related to the chassis (M, in lower frequencies) and to the wheel (m, in higher frequencies).
- The increase of the stiffness value displaces the resonance peak to higher frequencies.

3.1.6 Remarks on the Simplified Quarter-Car Model

In this section a simplified quarter-car model is considered. This kind of model is interesting to study since it partly reproduces the properties of the quarter-car model previously described while being much more simple. Such a model is given here as a didactic example but is rarely used in the literature for suspension analysis.

Definition 3.3 (*Simplified vertical passive quarter-car model*). *The simplified passive vertical quarter-car model, as depicted on Figure 3.5, is given by the following dynamical equation:*

$$M\ddot{z} = -k(z - z_r) - c(\dot{z} - \dot{z}_r) \tag{3.31}$$

where $k \in \mathbb{R}^+$ and $c \in \mathbb{R}^+$ are the linearized suspension and stiffness and damping coefficients, respectively. M is the chassis and wheel masses. z is the vertical chassis bounce displacement. Finally, z_r is the road vertical disturbance.

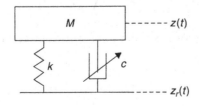

Figure 3.5: Simplified passive quarter-car model.

The main difference between Definition 3.3 and 3.1 is that the wheel dynamic is no longer described. As a consequence, the filtering effect (and its resonance peak) is no longer modeled. As an illustration, the following transfer function between the road and the chassis, $F_z(s)$, holds:

$$F_z(s) = \frac{cs + k}{Ms^2 + cs + k} \tag{3.32}$$

Compared to the quarter-car model, the system is of order two (while four for the complete). This property makes this example interesting and suitable for further analysis. As an illustration, natural frequency and damping ratio can be easily derived:

$$
\begin{aligned}
F_z(s) &= \frac{cs + k}{Ms^2 + cs + k} \\
&= \frac{\frac{c}{M}s + \frac{k}{M}}{s^2 + \frac{c}{M}s + \frac{k}{M}}
\end{aligned}
\tag{3.33}
$$

Then, by identifying the denominator with polynomial $s^2 + 2\xi\omega_n s + w_n^2$, it leads to:

$$w_n = \sqrt{\frac{k}{M}}$$

$$\xi = \frac{c}{2}\sqrt{\frac{1}{kM}} \tag{3.34}$$

And similarly to the quarter-car model, this model presents invariant points. As an illustration, the following property is given.

Property 3.4 (*Simplified quarter-car invariant points of $F_z(j\omega)$*).   *The transfer $F_{z_t - z_r}(j\omega)$ of the quarter-car model, as given in Definition 3.3, has two invariant points in $\omega \in \mathbb{R}^+$, defined as follows:*

$$
\begin{cases}
\tilde{\omega}_1 = 0 \\
\tilde{\omega}_2 = \sqrt{\frac{2k}{M}}
\end{cases}
\tag{3.35}
$$

Proof of Property 3.4:   The proof is similar to the previous one related to Property 3.1. Here the equation to be solved is:

$$\omega^2(M\omega^2 + 2k) = 0 \tag{3.36}$$

■

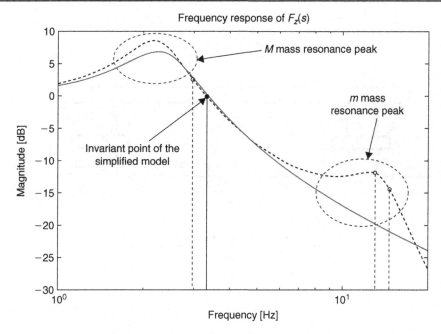

Figure 3.6: Frequency response $F_z(s)$: comparison between the quarter-car model (dashed line) and its simplified version (solid line) for $c = c_{min}$.

According to Figure 3.6, the simplified version of the quarter-car reproduces the resonance of the body mass quite well but the wheel resonance vanishes. Consequently, this kind of model may reasonably be used when focus is put on the chassis mass with respect to road unevenness, but is lacking in wheel dynamics.

3.2 Passive Vertical Half-Vehicle Models

The quarter-car studies are restricted to vertical behavior, thus vertical half-vehicle models are the natural extension of the vertical quarter-car model introduced in Section 3.1. Half-vehicle models simply involve an additional dynamic: e.g. the pitch (ϕ) or roll (θ) motion. Since these models are not used in this book, the analysis will be briefer and left to the reader's discretion. The presentation of these models is done for sake of completeness and focuses on nonlinear passive dynamical models. Readers may find more complete description and analysis in Dorling et al. (1995); Gáspár and Bokor (2006); Gáspár et al. (2004b, 2005); Sammier (2001); Sammier et al. (2000).

3.2.1 Pitch Oriented Model

The nonlinear model involving pitch dynamics, consists in linking two quarter-car models, as illustrated on Figure 3.7. Then, the associated definition is given as in Definition 3.4.

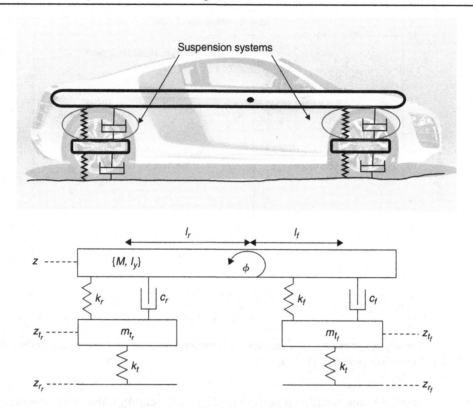

Figure 3.7: Half-car model (pitch oriented).

Definition 3.4 (*Passive vertical nonlinear dynamical half-car model – pitch oriented*). *The nonlinear vertical half-car pitch oriented model, as depicted on Figure 3.7, is given by the following dynamical equations:*

$$\begin{cases} M\ddot{z} = -k_f(z_f - z_{t_f}) - k_r(z_r - z_{t_r}) - c_f(\dot{z}_f - \dot{z}_{t_f}) - c_r(\dot{z}_r - \dot{z}_{t_r}) \\ m_{t_f}\ddot{z}_{t_f} = k_f(z_f - z_{t_f}) + c_f(\dot{z}_f - \dot{z}_{t_f}) - k_t(z_{t_f} - z_{r_f}) \\ m_{t_r}\ddot{z}_{t_r} = k_r(z_r - z_{t_r}) + c_r(\dot{z}_r - \dot{z}_{t_r}) - k_t(z_{t_r} - z_{r_r}) \\ I_y\ddot{\phi} = l_f\left(k_f(z_f - z_{t_f}) + c_f(\dot{z}_f - \dot{z}_{t_f})\right) - l_r\left(k_r(z_r - z_{t_r}) + c_r(\dot{z}_r - \dot{z}_{t_r})\right) + M_{dy} \end{cases}$$

(3.37)

$$\begin{cases} z_f = z + l_f\cos(\phi) \\ z_r = z - l_r\cos(\phi) \end{cases}$$

(3.38)

where $k_f \in \mathbb{R}^+$, $k_r \in \mathbb{R}^+$ and $k_t \in \mathbb{R}^+$ are the linearized front, rear suspension and tire stiffness coefficients respectively. $c_f \in \mathbb{R}^+$ and $c_r \in \mathbb{R}^+$ are the front and rear damping coefficients. M, m_{t_f} and m_{t_r} are the the chassis and front/rear wheel masses respectively. I_y is the pitch *inertia. l_f and l_r are respectively the distances between the center of gravity and the vehicle front and rear. z (resp. z_f and z_r), z_{t_f} and z_{t_r} are the vertical COG chassis (resp. front/rear) and front and rear wheel bounce displacements. Finally, z_{r_f}, z_{r_r} and M_{dy} are respectively the road front and rear displacements and the pitch moment, acting as disturbances.*

The reader can see that this model is simply the linking of two quarter-cars, with the addition of a new equation representing pitch dynamics ($\ddot{\phi}$) and a kinematic relation, defining front and rear position of the chassis (z_f and z_r). This kind of model is interesting when front/rear load transfer should be taken into consideration, i.e. in acceleration or braking situations.

Remark 3.2 (*About the roll oriented model*).   *Similarly to the pitch oriented Definition 3.4, the nonlinear roll oriented model can be derived in the same way, by replacing the pitch dynamic equation ϕ with a roll one θ, involving left and right suspensions and roll inertia. The roll model is usually used when load transfer between left and right of the vehicle are to be handled. It is typically the case in steering situations (see e.g. Gáspár et al., 2007; Poussot-Vassal et al., 2006; Sammier et al., 2000; Sampson and Cebon, 2003).*

The frequency responses are likely similar to the quarter-car model and the remarks for the quarter model also hold for these, more specifically:

- Both LPV (w.r.t. the spring stiffness) and LTI models can easily be derived from Definition 3.4.
- As for the vertical quarter-car model, the main nonlinearities are introduced by the suspension elements. However, kinematic relations also introduce nonlinear phenomena in these models.
- As for the quarter vertical model, the extended one (involving the longitudinal dimension) can easily be derived (see Section 3.4).

3.2.2 Numerical Discussion and Analysis

As for the quarter-car model, a numerical discussion and analysis is carried out and the pitch oriented half-vehicle model presented. Here, attention is given to the new variable and dynamic introduced in these models, namely, the pitch angle (ϕ) (see Figure 3.8). Additionally, a comparison between the vertical behavior with respect to road unevenness is shown with the quarter-car model (Figure 3.9). The models used in the following simulations are the ones described by Definition 3.4 with the automotive parameters given in Table 1.1 and with $c \in [c_{min}, c_{max}]$. Note that this pitch model is designed so that front and rear parameters are the same (i.e. $k_f = k_r$, $c_f = c_r$, $k_{t_f} = k_{t_r}$, $m_f = m_r$, $l_f = l_r$), then, the vehicle model is perfectly symmetric (which is obviously not true in reality). Indeed, in our specific case, asymmetrical modeling will not bring any interesting information, thus this assumption simplifies the analysis (for details, see Poussot-Vassal, 2008).

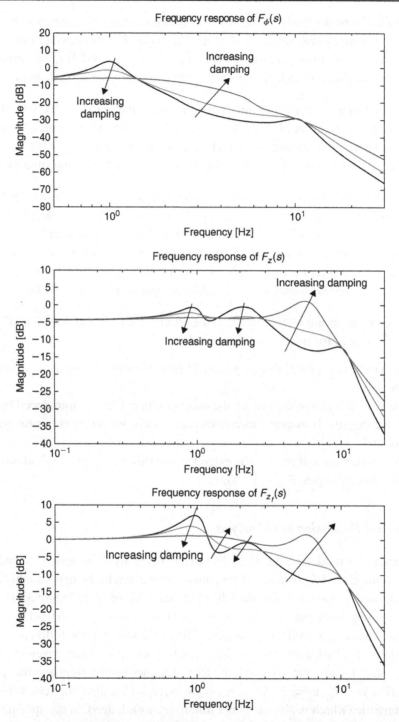

Figure 3.8: Bode diagram of the pitch at the center of gravity $F_\phi(s)$ (top), the bounce $F_z(s)$ at the center of gravity and of the front bounce $F_{z_f}(s)$ (bottom) of the pitch model for varying damping value c.

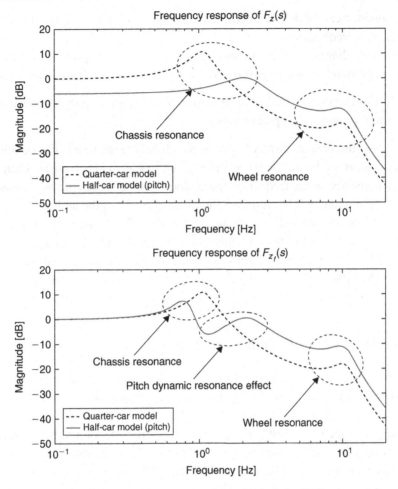

Figure 3.9: Bode diagrams of $F_z(s)$ and $F_{z_f}(s)$ for the half pitch (solid line) model, compared with for the quarter-car model (dashed line), for $c = c_{min}$.

First, Figure 3.8 presents the effect of the damping coefficient on the half-vehicle model on the dynamics at the center of gravity (COG) of the chassis. The following observations can be made:

- The pitch dynamic (ϕ) response with respect to the front road unevenness presents the same general characteristics as the chassis bounce of the quarter-car model presented in Section 3.1, namely, two peaks at specific frequencies, invariant points with respect to the damping coefficient, etc.
- The front bounce (z_f) dynamical response with respect to the front road unevenness presents a particular behavior. In addition to both classical peaks related to the chassis and wheel dynamics, an additional peak appears in between these at specific frequencies, caused by the pitch dynamic. One again, the increase of the damping value of the suspension system tends to remove the frequency oscillation.

- As for the quarter-car model, the increase of damping factor has as its principal effect the filtering of peak frequencies.
- Again, the Bode diagrams of the chassis and roll displacement present invariant points. These invariant points can easily be derived using the same analysis as in Section 3.1.

On Figure 3.9, behaviors of the half front dynamics are compared to the classical quarter-car model. Then, the following main points arise:

- By considering the bounce displacement of the chassis at the COG (Figure 3.9 – top), the static gain is clearly attenuated; this is mainly due to kinematic changes. Then, the peaks are simply translated in the frequency space, and the magnitudes of the resonance peaks are obviously attenuated. This is due to the filtering effect of the additional dynamic (i.e. the pitch). Indeed, the pitch dynamics add a filtering effect to the system, acting as load disturbances. This filtering effect attenuates the oscillation of the chassis.
- On Figure 3.9 – bottom, the front chassis bounce of the quarter and half-car models are compared. The main factor concerns the new frequency peak appearing between the chassis and wheel. This new resonance is due to the pitch dynamics.

3.3 Passive Vertical Full Vehicle Model

Here, attention is given to the full vertical vehicle model description which is the concatenation of the two previously introduced half models, including vertical, roll and pitch dynamics. First, assumptions under which the model is described are introduced, then, kinematic equations (due to the vehicle geometry) are provided, and finally, the dynamical equations are listed. Once again, this model is given for sake of completeness (more information may be found in Milliken and Milliken [1995] and Kiencke and Nielsen [2000]).

3.3.1 Assumptions and Kinematic Equations

Full vehicle modeling is not a simple task and involves many subsystems and coupled nonlinear dynamics. The following modeling assumptions have been considered:

- The kinematic effects due to suspension geometry are ignored (i.e. the suspensions only provide vertical forces to the chassis). Note that this assumption is always valid by an appropriate change of basis on the suspension system.
- The anti-roll bars are not considered (note that they play an important role for heavy vehicles, subject to important roll moments). Indeed, anti-roll bars provide an attenuation effect in the gain of the roll angle and on its cut-off frequency. This bar is increasingly controlled in global chassis control to avoid truck rollover (Gáspár and Bokor, 2006; Gáspár et al., 2004b,c; Sampson and Cebon, 2003).
- The vehicle chassis plane is considered parallel to the road while usually cars are bent over to improve air penetration and reduce aerodynamical resistance. Once again, the assumption is valid, if an adequate change of basis is considered.

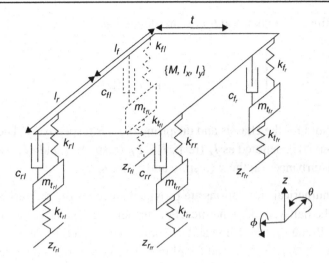

Figure 3.10: Full vertical vehicle model.

The kinematic equations are caused by the vehicle geometry. Each corner of the vehicle is identified with $\{i, j\}$ index, where $i = \{f, r\}$ holds for front/rear and $j = \{l, r\}$ for left/right. The chassis corners (i.e. positions and velocities of the dynamical part) are described as,

$$\begin{cases} z_{fl} = z + l_f \sin(\phi) - t \sin(\theta) \\ z_{fr} = z + l_f \sin(\phi) + t \sin(\theta) \\ z_{rl} = z - l_r \sin(\phi) - t \sin(\theta) \\ z_{rr} = z - l_r \sin(\phi) + t \sin(\theta) \\ \dot{z}_{fl} = \dot{z} + \dot{\phi} l_f \cos(\phi) - \dot{\theta} t \cos(\theta) \\ \dot{z}_{fr} = \dot{z} + \dot{\phi} l_f \cos(\phi) + \dot{\theta} t \cos(\theta) \\ \dot{z}_{rl} = \dot{z} - \dot{\phi} l_r \cos(\phi) - \dot{\theta} t \cos(\theta) \\ \dot{z}_{rr} = \dot{z} - \dot{\phi} l_r \cos(\phi) + \dot{\theta} t \cos(\theta) \end{cases} \tag{3.39}$$

where z is the center of gravity of the chassis mass, ϕ (resp. θ) is the pitch (resp. roll) angle of the chassis at the center of gravity. l_f, l_r and t define the vehicle geometrical properties (see Figure 3.10).

3.3.2 Full Vertical Dynamic Equations

The nonlinear vertical full car model, as depicted on Figure 3.10, is given by the following dynamical equations (3.40).

$$\begin{cases} M\ddot{z} = -F_{sz_{fl}} - F_{sz_{fr}} - F_{sz_{rl}} - F_{sz_{rr}} \\ m_{t_{ij}} \ddot{z}_{t_{ij}} = F_{sz_{ij}} - F_{tz_{ij}} \\ I_x \ddot{\theta} = (F_{sz_{rl}} - F_{sz_{rr}})t + (F_{sz_{fl}} - F_{sz_{fr}})t + M_{dx} \\ I_y \ddot{\phi} = (F_{sz_{rr}} + F_{sz_{rl}})l_r - (F_{sz_{fr}} + F_{sz_{fl}})l_f + M_{dy} \end{cases} \tag{3.40}$$

where the vertical tire and suspension forces are defined as

$$
\begin{cases}
F_{t_{ij}} = k_t (z_{t_{ij}} - z_{r_{ij}}) \\
F_{s_{ij}} = k(z_{ij} - z_{t_{ij}}) + c_{ij}(\dot{z}_{ij} - \dot{z}_{t_{ij}})
\end{cases}
\tag{3.41}
$$

where M and $m_{t_{ij}}$ hold for the chassis and unsprung masses respectively. The vehicle inertia in the x-axis (resp. y-axis) is denoted as I_x (resp. I_y). F_L (resp. M_{dx} and M_{dy}) are external forces (resp. moments) disturbances on the x (resp. y and z} axes.

In this section, no numerical simulations are provided since results are very similar to the ones obtained for the half-vehicle dynamics. For further analysis the reader is invited to refer to Andreasson and Bunte (2006); Chou and d'Andréa Novel (2005); Poussot-Vassal et al. (2008a, 2009), where this model is used for global chassis control using either active or semi-active suspensions.

3.4 Passive Extended Half-Vehicle Model

In this section, the extended half-vehicle model is introduced. The aim of this model is to extend the vertical half pitch oriented model described in Definition 3.4, by including the longitudinal dynamics, illustrating the existing link between the suspension system and this additional phenomenon.

Because of its complexity (nonlinear and of varying velocity dynamics), this model is not greatly used in the literature for control purposes, but is often used for validation. It is very important to understand the link between vertical behavior and longitudinal dynamics. This model helps us to illustrate the importance of the suspension system in critical driving situations; in some ways, it justifies the double comfort and road-handling objectives of the suspension system as described in Chapter 4.

3.4.1 Nonlinear Model

The nonlinear extended vehicle model is illustrated on Figure 3.11. Note that such a model completely describes (in a simplified version) the vertical and longitudinal dynamics of a motorcycle (for more details on motorcycle modeling, the reader should refer to the work of Tanelli [2007] and Corno [2009]).

Definition 3.5 (*Extended passive vertical nonlinear dynamical half-car model*). *The extended passive nonlinear vertical half-car pitch oriented model, as depicted on Figure 3.11,*

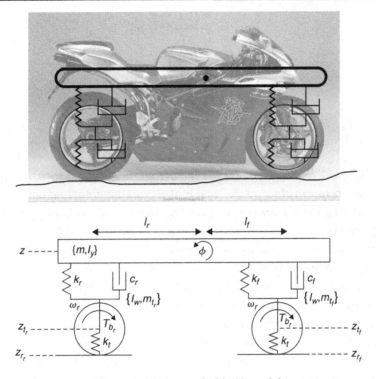

Figure 3.11: Extended half-model.

is given by the following dynamical equations:

$$\begin{cases}
M\ddot{z} = -k_f(z_f - z_{t_f}) - k_r(z_r - z_{t_r}) - c_f(\dot{z}_f - \dot{z}_{t_f}) - c_r(\dot{z}_r - \dot{z}_{t_r}) \\[4pt]
m_{t_f}\ddot{z}_{t_f} = k_f(z_f - z_{t_f}) + c_f(\dot{z}_f - \dot{z}_{t_f}) - k_t(z_{t_f} - z_{r_f}) \\[4pt]
m_{t_r}\ddot{z}_{t_r} = k_r(z_r - z_{t_r}) + c_r(\dot{z}_r - \dot{z}_{t_r}) - k_t(z_{t_r} - z_{r_r}) \\[4pt]
I_y\ddot{\phi} = l_f\left(k_f(z_f - z_{t_f}) + c_f(\dot{z}_f - \dot{z}_{t_f})\right) - l_r\left(k_r(z_r - z_{t_r}) + c_r(\dot{z}_r - \dot{z}_{t_r})\right) + M_{dy} \\[4pt]
\dot{\lambda}_f = -\dfrac{1}{v}\left(\dfrac{1-\lambda_f}{m_f} - \dfrac{R^2}{I_w}\right)F_{tx}(\mu_f, \lambda_f, F_{n_f}) + \dfrac{R}{vI_w}T_{b_f} \\[4pt]
\dot{\lambda}_r = -\dfrac{1}{v}\left(\dfrac{1-\lambda_r}{m_r} - \dfrac{R^2}{I_w}\right)F_{tx}(\mu_f, \lambda_r, F_{n_r}) + \dfrac{R}{vI_w}T_{b_r} \\[4pt]
I_w\dot{\omega}_f = R_t F_{tx}(\mu_f, \lambda_f, F_{n_f}) - T_{b_f} \\[4pt]
I_w\dot{\omega}_r = R_t F_{tx}(\mu_r, \lambda_r, F_{n_r}) - T_{b_r} \\[4pt]
M\dot{v} = -F_{tx}(\mu_f, \lambda_f, F_{n_f}) - F_{tx}(\mu_r, \lambda_r, F_{n_r})
\end{cases} \tag{3.42}$$

$$\begin{cases}
z_f = z + l_f\cos(\phi) \\
z_r = z - l_r\cos(\phi)
\end{cases} \tag{3.43}$$

$$\begin{cases} m_f = \dfrac{l_f}{l_f + l_r} M + m_{t_f} \\[3mm] m_r = \dfrac{l_r}{l_f + l_r} M + m_{t_r} \end{cases} \tag{3.44}$$

$$\begin{cases} F_{n_f} = -m_f g + k_f (z_f - z_{t_f}) + c_f (\dot{z}_f - \dot{z}_{t_f}) - k_t (z_{t_f} - z_{r_f}) \\[2mm] F_{n_r} = -m_r g + k_r (z_r - z_{t_r}) + c_r (\dot{z}_r - \dot{z}_{t_r}) - k_t (z_{t_r} - z_{r_r}) \end{cases} \tag{3.45}$$

where $k_f \in \mathbb{R}^+$, $k_r \in \mathbb{R}^+$ and $k_t \in \mathbb{R}^+$ are the linearized front, rear suspension and tire stiffness coefficients respectively. $c_f \in \mathbb{R}^+$ and $c_r \in \mathbb{R}^+$ are the front and rear damping vertical coefficients. T_{b_f} and T_{b_r} are the braking torques (not used at all is this study, i.e. $= 0$). l_f and l_r are respectively the distances between the center of gravity and the vehicle front and rear. M, m_{t_f} and m_{t_r} are the the chassis and front/rear wheel masses respectively. g is the gravitational constant. I_y and I_w are the pitch and wheel inertia respectively. R_t is the nominal tire radius length. z (resp. z_f and z_r), z_{t_f} and z_{t_r} are the vertical chassis (resp. front/rear) and front and rear wheel bounce displacements. ω_f, λ_f (resp. ω_r λ_r) are the front (resp. rear) rotational wheel velocity and slip. v is the vehicle longitudinal velocity. z_{r_f}, z_{r_r} and M_{dy} are respectively the road front and rear displacements and pitch moment disturbances. F_{tx} defines the longitudinal friction force, which is function of λ_f, λ_r, μ_f and μ_r, the front and rear road adhesion surface and normal front and rear loads are defined by F_{n_f} and F_{n_r}.

From this model definition, it is obvious that the coupling phenomenon i.e. the link between vertical $\{z, z_t\}$ and longitudinal $\{\lambda, \omega, v\}$ dynamics is the normal front (F_{n_f}) and rear (F_{n_r}) loads which are a function of the suspension force. The load control (though suspension control) is a real challenge to enhance longitudinal performances also. This model is increasingly studied in the literature since it can connect the work of both brake/traction and suspension control communities, but still a few results are available. Note that the motorcycle community is very involved in such models since the pitch dynamic is more accentuated in two-wheeled vehicles than in four-wheeled ones (see e.g. recent works of Corno [2009]; Poussot-Vassal [2007]).

3.5 Semi-Active Vertical Quarter-Car Model

Here, the focus is placed on the semi-active model which will be used throughout the book. The models presented hereafter extend the one previously introduced.

3.5.1 Nonlinear and LTI Models

As illustrated on Figure 3.12 (right) and suggested in Chapter 2, when controlled suspension is considered, the passive damper with frozen damping c (i.e. non-controllable), is replaced by a controllable shock absorber with adjustable damping $c(t)$. In this case, F_d denotes the force

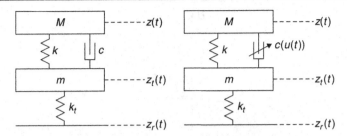

Figure 3.12: Passive (left) and semi-active (right) quarter-car models.

provided by the controllable device, according to the damping request c_{in}. As an illustration, if a magnetorheological damper is considered, the damping request is driven by the mean of an adjustable current $I(t)$ (as described in Chapter 2).

Definition 3.6 (*Dynamical nonlinear semi-active vertical quarter-car model*). *According to first-order nonlinear damper model (2.23), the dynamical semi-active nonlinear vertical quarter-car model, as depicted on Figure 3.12 (right), is given by the following dynamical equations:*

$$\begin{cases} M\ddot{z} = -k(z - z_t) - c(\dot{z} - \dot{z}_t) \\ m\ddot{z}_t = k(z - z_t) + c(\dot{z} - \dot{z}_t) - k_t(z_t - z_r) \\ \dot{c} = -\beta c + \beta c_{in} \end{cases} \quad (3.46)$$

where $k \in \mathbb{R}^+$ and $k_t \in \mathbb{R}^+$ are the linearized suspension and tire stiffness coefficients, respectively. c and c_{in} are the actual and the requested damping coefficient, respectively. β stands for the controlled damper actuator bandwidth. M and m are the chassis and wheel masses, respectively. z and z_t are the vertical chassis and wheel bounce displacements. Finally, z_r is the road vertical disturbance.

As underlined in the semi-active model Definition 3.6, the nonlinear phenomena come from the equation structure, and more specifically from the coupling between two system states, namely c and $\dot{z} - \dot{z}_t$. Consequently, it is not possible to derive an LTI semi-active quarter-car model by applying classical linearization techniques as in Definition 3.1. But still it is possible to define a LTI control oriented model, which will then be used for control design, as follows:

Definition 3.7 (*LTI control oriented quarter-car model*). *The LTI control oriented quarter-car model, denoted as $\Sigma_c(c^0)$, is given by:*

$$\Sigma_c(c^0) := \begin{cases} M\ddot{z} = -k(z - z_t) - c^0(\dot{z} - \dot{z}_t) - F_d \\ m\ddot{z}_t = k(z - z_t) + c^0(\dot{z} - \dot{z}_t) + F_d - k_t(z_t - z_r) \\ \dot{F}_d = -\beta F_d + \beta u \end{cases} \quad (3.47)$$

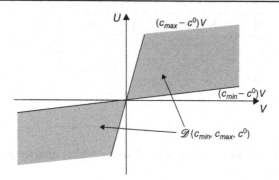

Figure 3.13: Dissipative domain $\mathscr{D}\left(c_{min}, c_{max}, c^0\right)$ graphical illustration.

where $k \in \mathbb{R}^+$ and $k_t \in \mathbb{R}^+$ are the linearized suspension and tire stiffness coefficients, respectively. $c^0 \in \mathbb{R}^+$ is the linearized nominal damping coefficient of the suspension system. F_d and u are the actual and the requested damping forces, respectively. β stands for the controlled damper actuator cut-off frequency. M and m are the chassis and wheel masses respectively. z, z_t are the vertical chassis and wheel bounce displacements. Finally, z_r is the road vertical disturbance.

Then, to fit the semi-active control constraints, the additional force reference u should satisfy the linear dissipative constraints $u \in \mathscr{D}(c_{min}, c_{max}, c^0) \subseteq \mathbb{R}$, where the dissipative $\mathscr{D}(c_{min}, c_{max}, c^0)$ set is defined as follows (see also Figure 3.13):

$$\mathscr{D}(c_{min}, c_{max}, c^0) := \left\{ U \in \mathbb{R} | \forall V \in \mathbb{R} : (U - (c_{max} - c^0)V)((c_{min} - c^0)V - U) \geq 0 \right\}$$

(3.48)

where c_{min} and c_{max} are the minimal and maximal damping factors of the considered controlled damper, normalized around c^0.

Note that,

$$u \in \mathscr{D}(c_{min}, c_{max}, c^0) \Leftrightarrow F_d \in \mathscr{D}(c_{min}, c_{max}, c^0)$$

(3.49)

Hence, if the reference force provided by the controller is "semi-active", then it will be achieved by the considered actuator.

Remark 3.3 (*About the c^0 parameter*). *The c^0 parameter, appearing in the LTI control oriented quarter-car Definition 3.7, may be omitted (i.e. $c^0 = 0$), but, selecting a non-null c^0 parameter is very convenient for numerical issues. Indeed, in some simulations and for controller synthesis purposes, a nominal damping factor should be introduced in order to avoid oscillatory behaviors and numerical explosion due to poles in limit of stability (see Chapter 5 and poles displacement as a function of the damping value for a deeper illustration and explanation of this point).*

Then, similarly to the passive case, the associated LTI state space representation can be derived as follows:

$$\dot{x} = A(c^0)x + B \begin{bmatrix} z_r \\ u \end{bmatrix} \tag{3.50}$$

or equivalently,

$$\begin{bmatrix} \ddot{z} \\ \dot{z} \\ \ddot{z}_t \\ \dot{z}_t \\ \dot{F}_d \end{bmatrix} = \begin{bmatrix} \frac{-c^0}{M} & \frac{-k}{M} & \frac{c^0}{M} & \frac{k}{M} & \frac{-1}{M} \\ 1 & 0 & 0 & 0 & 0 \\ \frac{c^0}{m} & \frac{k}{m} & \frac{-c^0}{m} & \frac{-k-k_t}{m} & \frac{1}{m} \\ 0 & 0 & 1 & 0 & 0 \\ 0 & 0 & 0 & 0 & -\beta \end{bmatrix} \begin{bmatrix} \dot{z} \\ z \\ \dot{z}_t \\ z_t \\ F_d \end{bmatrix} + \begin{bmatrix} 0 & 0 \\ 0 & 0 \\ \frac{k_t}{m} & 0 \\ 0 & 0 \\ 0 & \beta \end{bmatrix} \begin{bmatrix} z_r \\ u \end{bmatrix} \tag{3.51}$$

Note that the main difference between the passive and the controlled systems is the additional state F_d, defining the control input. The main point to keep in mind when defining the LTI controlled quarter-car model is the force limitation as given on Figure 3.13. Otherwise, the controlled quarter-car model may become active.

From now on, Definitions 3.1 and 3.7, will be the ones to refer to when passive or semi-active quarter-vehicle models are considered for control, unless explicitly specified. No specific analysis of the semi-active quarter-car model is done since it would be very similar to the passive one and since the most interesting analysis to carry out on this kind of model is closely related to the control strategy (see the following chapters).

3.5.2 Toward LPV Models...

As previously explained in Definition 2.1, the spring model may be nonlinear, resulting in a model where the stiffness k is varying in a closed set (i.e. $\underline{k} \leq k \leq \overline{k}$) according to the suspension deflection z_{def}. Thanks to the LPV (linear parameter varying) modeling theory, it is possible to take these nonlinearities into account as proposed in Zin et al. (2008). For more insight into this kind of modeling, the reader is encouraged to read the excellent works of Bruzelius (2004) and Toth (2008) (see also works on LPV control systems of Balas et al. [2003]; Gáspár and Bokor [2006]; Gáspár et al. [2004a; 2004c]; Shamma and Athans [1991, 1992]).

As a light introduction and as an illustration, let us consider that the spring force is a nonlinear function of the deflection $(z - z_t)$, then, an LPV (indeed quasi-LPV) model formulation can be derived by selecting as the varying parameters the nonlinear stiffness coefficients of the suspension (see e.g. Zin et al., 2008). A qLPV model can then be described by the following parameter dependent dynamical equations:

$$\begin{cases} M\ddot{z} = -k(.)(z - z_t) - c(\dot{z} - \dot{z}_t) - F_d \\ m\ddot{z}_t = k(.)(z - z_t) + c(\dot{z} - \dot{z}_t) + F_d - k_t(z_t - z_r) \end{cases} \tag{3.52}$$

where $k(.)$ is the varying nonlinear stiffness coefficient of the suspension system, obtained by inverting the nonlinear characteristic of the spring curves. Consequently, the qLPV associated "state-space" representation can be given as:

$$
\begin{bmatrix} \ddot{z} \\ \dot{z} \\ \ddot{z}_t \\ \dot{z}_t \\ \dot{F}_d \end{bmatrix} = \begin{bmatrix} \frac{-c}{M} & \frac{-k(.)}{M} & \frac{c}{M} & \frac{k(.)}{M} \\ 1 & 0 & 0 & 0 \\ \frac{c}{m} & \frac{k(.)}{m} & \frac{-c}{m} & \frac{-k(.)-k_t}{m} \\ 0 & 0 & 1 & 0 \end{bmatrix} \begin{bmatrix} \dot{z} \\ z \\ \dot{z}_t \\ z_t \end{bmatrix} + \begin{bmatrix} 0 & -\frac{1}{M} \\ 0 & 0 \\ \frac{k_t}{m} & \frac{1}{m} \\ 0 & 0 \end{bmatrix} \begin{bmatrix} z_r \\ F_d \end{bmatrix}
\tag{3.53}
$$

In this case, it is straightforward that:

- The parameter dependency enters in a linear way in the dynamical state matrix, i.e. the system has the following general form:

$$
\dot{x} = (A^0 + A^1 k(.))x + B z_r
\tag{3.54}
$$

  where A^0 stands for the constant elements of the A matrix, while A^1 stands for the k-dependent dynamical matrix.
- The parameter $k(.)$ is function of the state variables, i.e. $k(.)$ is function of $z_{def} = z - z_t$.

This kind of modeling should then be used to design controllers gain-scheduled by these coefficients to improve performance and robustness (see e.g. Poussot-Vassal et al., 2008b; Zin et al., 2008).

3.6 Conclusions

In this chapter, the most common models used to analyze the suspension system and design associated controllers are presented. This presentation is done by increasing the model complexity, starting from the simple quarter-car model involving vertical dynamics only, towards the full vertical model involving roll and pitch dynamics also. In addition, the extended half model is also presented, illustrating the link between the vertical forces (hence the suspension actuator) with the longitudinal dynamics. The chapter closes with the definition of the semi-active quarter-car models which will be extensively used throughout the book.

The chapter provides a detailed analysis of the quarter-car model, with both analytic and numerical developments. Then, more complex models are presented for the sake of completeness, but with a shortened analysis. From the analysis point of view, efforts are focused on the damping element and on its influence on the vehicle dynamics with respect to road unevenness. The main objective of this chapter was to introduce in a comprehensive manner the commonly used models in both literature and industry for analyzing the suspension system. Properties and analysis can be extended, but together with Chapters 1 and 2, this first overview provides the reader with the necessary insight and background to understand the challenges and limitations in controlled dampers. We aimed to introduce the model notations and make the reader familiar with the frequency representations largely used in this book.

The next chapter is devoted to the definition of the signals of interest and metrics to characterize the suspension properties (crucial for control design), i.e. comfort or road-holding. Notations and notions given in the present chapter will be, from now on, assumed to be known and used without any specific recall.

CHAPTER 4

Methodology of Analysis for Automotive Suspensions

System performances evaluation is a key issue in control theory and applications as well as in almost all the engineering fields. Indeed, for any engineering problem, one aims at evaluating how efficient a system is, and, if possible, to set up some measure to evaluate in a systematic and objective way such an efficiency. In control theory, measures are widely exploited, see e.g. Boyd et al. (1994); Dorf and Bishop (2001); Zhou et al. (1996) (e.g. Time response or $\mathcal{H}_\infty$, $\mathcal{H}_2$ norms...). In control applications, each system may have its own performance measure (e.g. braking distance for an ABS, energy consumption for a motor...).

Regarding the suspension systems (either passive, semi-active or active), it is of great importance to have specific evaluation tools to measure the efficiency of a new structure, a novel control algorithm... in order to evaluate the improvements brought with respect to a nominal reference system. More specifically, when dealing with suspension systems, the two main aspects of interest (for both academic and industry applications) are concerned with:

- The comfort characteristics.
- The road-holding (or handling) characteristics.

This chapter is devoted to the presentation of time and frequency domains tools used both in the academic and industry literature to characterize and analyze performance of passive, semi-active and active suspension systems. This chapter does not claim to present all the aspects of comfort and road-holding performance analysis, but still provides general methodologies and metrics to evaluate a suspension system (controlled or not controlled). Note that these performances will then be used to evaluate the developed control strategies (see Chapter 5 to Appendix B).

The chapter is structured as follows: in Section 4.1, after a brief description of human body comfort and vehicle road-holding specifications, the introduction of the two points of interest to characterize comfort and road-holding properties is made, based on the quarter-car model. Then, the frequency domain performance evaluation procedure is presented in Section 4.2, together with the performance indexes aiming at measuring how comfortable/road-holding a vehicle is. Time domain performance evaluation algorithms are described in Section 4.3.

Conclusions and discussions are given in Section 4.4. All results and notions are illustrated and discussed through numerical simulations performed on the quarter-car model given in Definition 3.1, using the motorcycle parameters of Table 1.2.

4.1 Human Body Comfort and Handling Specifications

In the automotive field, vehicle and human body performance specifications are subjects where many works have been published (and still are open research areas covering different engineering domains) (Gillespie, 1992). Briefly speaking, to define these performances, automotive engineers have first to formulate requirements in "classical words" (e.g. not engineering terminology), then, to turn them into mathematical expressions and introduce consistent and repetitive metrics to quantify these characteristics. Usually, the first step is straightforward since it comes from driver feelings and expectation. The second step is usually much harder since it is very dependent on the engineering general approach (time, frequency...), and may turn out to be non-representative to all drivers. In this book, some metrics for performance evaluation have been selected according to the authors, general observations and discussions with automotive experts (from both industry and academy). The authors stress that other relevant metrics should also be used and may be preferred according to the reader, but, since the proposed ones are derived in both time and frequency domains, they are representative of a large spectrum of behaviors and can be considered as reliable and consistent for our purpose. Moreover, even if other metrics may be introduced, the reader should notice that the interest of the proposed metrics rely on their simplicity and complementarity, reflecting in a neat way the semi-active suspension control trade-off.

Since focus is on the controlled damper actuator, it is worth noting that in this book, the main dynamics under consideration are the vertical ones. As a consequence, the model used in this chapter is the classical quarter-car model. Suspension characteristics also affect other ground vehicle dynamics such as roll, pitch, longitudinal, but also yaw, therefore the damping control is also a matter of global chassis control improvement (see e.g. Poussot-Vassal et al., 2009; Zin, 2005).

4.1.1 Comfort Specifications

When comfort characteristics are considered, it mainly concerns the passenger's comfort. This very subjective feeling is a combination of different factors such as:

- The driver's state (e.g. feelings, age, health, general abilities etc.) and environment (e.g. weather) which are not controllable (at least with semi-active suspensions).
- The chassis vibrations, noise, etc. which are controllable by an adequate suspension and control law design.

From a general viewpoint, comfort feeling, characterized by the road unevenness filtering, has an impact on the driver reaction time, accuracy, situation evaluation and decision abilities, which makes this objective particularly active in the automotive community (see Chapter 6 and Fischer and Isermann, 2003).

Since the driver comfort study is a complex and subjective concept, an entire chapter may be required to describe driver vibration models, sensitive zones, etc. Roughly speaking, in the literature, comfort analysis is mainly treated as non-comfort analysis; such studies are usually held by modeling the human body as a complex system composed of masses linked by spring and damper elements, modeling the muscles (Girardin et al., 2006). Then, according to this kind of model, studies indicate some sensitive frequency zones (related e.g. to the heart, the head, etc.) having resonance or gain amplifications around some specific frequencies according to different disturbances (such as the steering wheel vibrations or the road irregularities).

In this work, human body sensitive functions are neither characterized nor directly discussed, but evaluated through the chassis analysis. Indeed, in this book framework, comfort feeling analysis is performed by analyzing some specific frequencies of the vertical behavior of the quarter-vehicle model (instead of vehicle passenger directly). This simplification allows us to first avoid human body modeling and uncertain parametrization and secondly to reduce the number of variables for comfort study. Therefore, the focus will be on the analysis of simpler variables behavior with respect to road unevenness, such as vertical acceleration ($\ddot{z}$) and displacement (z) of the chassis. Consequently, from now on, an improvement on these variables will imply passenger comfort improvement. Mathematically, the objective is simply:

$$\min \ddot{z}(t) = \min z(t) \tag{4.1}$$

4.1.2 Road-Holding Specifications

Compared to the human comfort, road-holding is a vehicle property which characterizes the ability of the vehicle (and more specifically, of the wheels) to keep contact with the road and maximize wheel tracking to road unevenness, i.e. to filter the wheel trepidations generated by road irregularities, and, to guarantee road contact whatever the road profile and load transfer situations (e.g. in difficult cornering situations). This contact property is essential since wheel load is strongly related to longitudinal and lateral force descriptions which are essential in all vehicle dynamics (especially longitudinal, lateral, yaw dynamics, see Chapter 3). To illustrate this last point, let us first describe, in a very simplified manner, the longitudinal (F_{tx}) lateral (F_{ty}) forces of each tire as follows:

$$F_{tx} = F_n \mu_x(\lambda, \vartheta)$$
$$F_{ty} = F_n \mu_y(\beta, \vartheta) \tag{4.2}$$

where μ_x and μ_y are the nonlinear functions, dependent on λ, β, ϑ, respectively the slip ratio, the slip angle and the road roughness characteristics, describing the longitudinal and lateral tire forces at the road contact point. These functions are highly nonlinear and complex to define; in fact, they will not be treated in this book, and assumed as constant (for more detail, the reader is encourage to refer to Canudas et al. (2003); Kiencke and Nielsen (2000); Tanelli (2007); Velenis et al. (2005)). These forces are also affine functions of F_n, the normal load, defined as:

$$F_n = (M + m)g - k_t(z_t - z_r) \tag{4.3}$$

where M and m are the chassis and wheel masses, k_t is the vertical tire stiffness characteristic.

Since the aim of the tire is to provide maximal longitudinal and lateral forces (for acceleration, braking and steering maneuver), it is obvious that the normal load F_n should be maximized. The handling objective consists in maximizing the vertical load. Therefore, since $(M + m)g > 0$ and $k_t > 0$, handling objective consists in:

$$\min(z_t(t) - z_r(t)) \tag{4.4}$$

where $z_t(t) - z_r(t)$ is the tire deflection. Once again, road-holding is a very critical performance and is very complex to describe and characterize using the simple models derived in Chapter 3. To be fully complete, the quarter-car model should include the wheel/road contact loss and a more complex tire model. For the sake of "simplicity" in this book, the road-holding characterization is only done by considering the tire deflection ($z_{def_t} = z_t - z_r$) minimization problem.

4.1.3 Suspension Technological Limitations: End-Stop

Additionally to both previous performance definitions, characterizing comfort and road-holding, it is also important, when evaluating a suspension system and its associated control algorithm, to take into account the static suspension stroke limitations $z_{def} = z - z_t$ which should always remain in between the limitations defined by the technology, i.e.

$$z_{def} \in \left[z_{def}^{min}, z_{def}^{max} \right] \tag{4.5}$$

where z_{def}^{min} and z_{def}^{max} and the suspensions end-stop, or deflection limits. This constraint is very important for practical applications, in order to preserve the mechanical suspension system. From a control engineer point of view, such a static state constraint is very hard to take into account (since it involves time constraints). But fortunately, since these limitations should not occur in classical driving situations, in practice, stroke limitations should not be directly handled by the controller. Therefore, in our application, and as explained in Chapter 3 (see also Poussot-Vassal, 2008) the suspension system is linearized and road disturbances are considered to be small. In fact, the limitations are not considered.

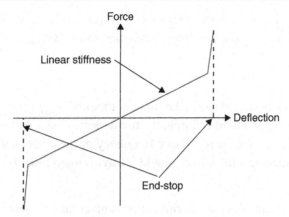

Figure 4.1: Nonlinear suspension stiffness and stroke limitations.

To illustrate this point, let us simply consider Figure 4.1 which represents a classical suspension stiffness as a function of the deflection.

Then, according to Figure 4.1, our main objective is to consider that the operating conditions remain in the linear zone.

4.1.4 Quarter-Car Performance Specifications and Signals of Interest

Considering the previous remarks, and considering the quarter-car model given in Definition 3.1, the following signals are considered for performance analysis and characterization of a suspension system:

- To evaluate the comfort, the vertical displacement z (or acceleration $\ddot{z}$) of the chassis is analyzed.
- To evaluate the road-holding, the tire deflection (z_{def_t}) is analyzed.
- Eventually, the deflection limits (z_{def}^{min} and z_{def}^{max}) could be analyzed. Note that in this book, this point will not be dealt with, since it involves nonlinear suspension spring description.

Figure 4.2 recalls the frequency responses of both main transfer functions under discussion in this book, namely, F_z and $F_{z_{def_t}}$. Then let us define the comfort and road-holding performances as follows.

4.1.4.1 Comfort

The vibration isolation between [0; 20]Hz is evaluated by the transfer function $F_z(s)$. Ideally, the vertical displacement of the chassis should be the same as the one of the road for low frequencies (lower than around 1 Hz) and attenuated for high frequencies (higher than 1 Hz), in order to filter undesirable vibrations. In practice, for low disturbances ($z_r < 3$ cm), the maximal gain occurring between 1 and 5 Hz of $F_z(s)$ has to be lower than 1.8, and over this frequency, road disturbances should be filtered (see Sammier, 2001).

4.1.4.2 Road-Holding

As indicated before, it is evaluated using the transfer function $F_{z_{def_t}}(s)$ over the frequency range of [0; 30]Hz. For a good road-holding, the tire deflection should be attenuated for low frequencies, and filtered around the resonance frequency of the wheel and over. Moreover, especially for high frequencies, the wheel should always remain linked (i.e. in contact) to the road.

Note that these objectives are consistent with the ones given in previous suspension analysis works of Gillespie (1992) and Sammier et al. (2003). On Figure 4.2, the comfort and road-holding performances objectives are illustrated, considering the passive quarter-car model using the motorcycle parameters given in Table 1.2.

4.1.4.3 Deflection

The deflection signal should remain in the linear zone of the suspension system to avoid the end-stop zone (leading to discontinuities). This consideration is very important especially when the comfort objective is to be reached.

Note that in this study, frequency limits will be fixed to 30 Hz. This choice is justified by the fact that, practically, higher frequencies are filtered by the vehicle mechanical elements. Therefore, it is not of great importance to extend the analysis over that frequency. Moreover, according to the quarter-car definition given in Chapter 3, this frequency limit is enough to describe all the interesting behaviors (i.e. the two masses resonance peaks, the static gain, the attenuation area, etc.).

4.2 Frequency Domain Performance Evaluations

In the previous section, comfort and road-holding specifications were presented on a quarter-car vehicle, using "classical words". This section is concerned with the description of two fundamental tools used throughout the book, namely: the frequency response (FR) and the performance indexes. Let us now introduce the frequency domain based metrics used to mathematically characterize the suspension.

4.2.1 Nonlinear Frequency Response Computation

In order to evaluate the semi-active suspension performances, the frequency response (FR) of the system is evaluated. Since suspension models and/or controlled control algorithms may be nonlinear, the closed-loop often turns to be nonlinear itself (see e.g. Aubouet et al., 2009;

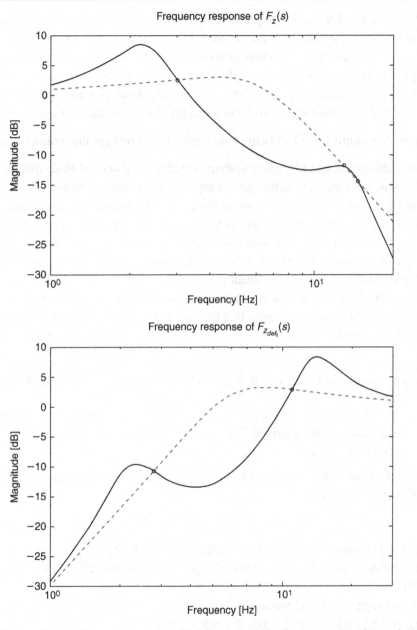

Figure 4.2: Illustration of the performance objectives on Bode diagrams. Comfort oriented diagram F_z (top) and Road-holding oriented diagram $F_{z_{def_t}}$ (bottom). Solid line: c_{min}, Dashed: c_{max}.

Poussot-Vassal et al., 2008c and Chapters 6–8 and Appendix A and B). Therefore, Bode diagrams cannot be computed anymore. Consequently, two simple discrete algorithms to compute frequency responses of nonlinear systems are proposed here. These algorithms allow, under some assumptions, to analyze the frequency behavior of nonlinear systems.

In the first one (Algorithm 1), the frequency response is evaluated frequency by frequency while the second one (Algorithm 2) excites multiple frequencies at the same time. For a linear system description, these two algorithms provide identical results, while for nonlinear systems, they may present some differences (but still provide similar results if signals are persistent enough). Note that the frequency representation is widely developed and used in both academic and industrial research since it provides a rich information set of the suspension performances.

According to Algorithm 1, the following comments should be kept into consideration:

- Applying this procedure to a linear system provides the so-called Bode diagram.
- The P parameter aims at avoiding gain computation on transient behaviors of the nonlinear system. By fixing $P = 10$ one computes the gain in the sinusoidal "steady state" behavior.
- The interest in this FR computation (which is not representative of a real road profile disturbance) is that it allows us to derive a Bode like diagram for nonlinear systems. In this book, this procedure is largely used to plot the frequency behavior of a suspension system controlled (or not) through a nonlinear control law.
- Since the FR algorithm aims at characterizing the frequency response of nonlinear systems, it is also important to note that the amplitude of the sinusoid disturbance will modify the frequency response. Consequently, nonlinear analysis should also be carried

---

Algorithm 1 Nonlinear Frequency Response Computation $\tilde{F}$ – FR (Single Frequency)

To compute the approximate frequency response of $\tilde{F}_z(f)$ – comfort characteristic – and $\tilde{F}_{def_t}(f)$ – road-holding characteristic – of a (controlled) nonlinear suspension system, the following procedure is applied:

1. A sinusoidal road disturbance $z_r(t)$ feeds the input of the nonlinear quarter vehicle models, over P periods such that:

$$z_r(t) = A \sin(2\pi f t) \tag{4.6}$$

   where $A \in [1; 5]$cm, $f \in [\underline{f}; \overline{f}] \subseteq [1; 30]$Hz and $t \in [P/\overline{f}; P/\underline{f}]$s, where $\{P \in \mathbb{Z} | P > 1\}$ is the number of periods of the sinusoid feeding the system. Practically one may choose $P = 10$.

2. The output signals $y(t)$ are measured. Here, $y(t) = z(t)$ (vertical suspended mass bounce) or $\ddot{z}(t)$, and $z_{def_t}(t) = z_t(t) - z_r(t)$ (tire deflection).

3. For each signal the corresponding spectrum $Y(f)$ of $y(t)$ (and $U(f)$ of $z_r(t)$) is computed (by means of a discrete Fourier Transform).

4. The power spectral density of $Y(f)$ and $U(f)$ signals are computed; denoted as $G_y(f)$ and $G_u(f)$.

5. For each output signal of interest, the variance gain is computed as: $\tilde{F}(f) = G_y(f)/G_u(f)$. In our case, $\tilde{F}_z(f)$ and $\tilde{F}_{z_{def_t}}(f)$ signals are obtained.

---

through varying amplitudes. For the sake of readability, in this book, the disturbance amplitude will often be reduced to one single amplitude where the suspension stays in its linear working point (i.e. $A = 1\,\text{cm}$ in Algorithm 1). Remember that on a real suspension, nonlinearities will modify the dynamics (e.g. hysteresis phenomena, which are often present in magnetorheological dampers, see Chapter 2).

Similarly to Algorithm 1, another approach can be used to derive the frequency response of a nonlinear system. This approach consists in exciting the system at multiple frequencies at the same time.

Algorithm 2   Nonlinear Frequency Response Computation F – FR (Multiple Frequency)

To compute the approximate frequency response of $F_z(f)$ – comfort characteristic – and $F_{def_t}(f)$ – road-holding characteristic – of a (controlled) nonlinear suspension system, the following procedure is applied:

1. A broad-band white noise limited bandwidth road disturbance $z_r(t)$ (of 0 mean and varying between z_r^{min} and z_r^{max}) feeds the input of the nonlinear quarter-vehicle models, for a duration of T seconds such that:

$$z_r(t) = N(0, \sigma) \tag{4.7}$$

    where $N(0, \sigma)$ defines the white noise signal of zero mean and variance σ.
2. The output signals $y(t)$ are measured. Here, $y(t) = z(t)$ (vertical suspended mass bounce) or $\ddot{z}(t)$, and $z_{def_t}(t) = z_t(t) - z_r(t)$ (tire deflection).
3. For each signal the corresponding spectrum $Y(f)$ of $y(t)$ (and $U(f)$ of $z_r(t)$) is computed (by mean of a discrete Fourier Transform).
4. The power spectral density of $Y(f)$ and $U(f)$ signals are computed; denoted as $G_y(f)$ and $G_u(f)$.
5. For each output signal of interest, the variance gain is computed as: $F(f) = G_y(f)/G_u(f)$. In our case, $F_z(f)$ and $F_{z_{def_t}}(f)$ signals are obtained.

When using this algorithm, it is important to simulate the nonlinear system over a long duration in order to provide a persistent excitation to the system (e.g. $T = 60$) and to remove the 5–10 first seconds for the variance gain computation in order to avoid transient behavior, non-representative of the real system.

4.2.2 Performance Index Computation

According to the previously introduced FR algorithm computations and signal of interest description, let us now introduce the so-called metrics, aiming at characterizing the comfort and road-holding performances. These suspension performance measures (or indexes) are defined as follows (note that two definitions are given, associated to each FR algorithm).

Definition 4.1 (*Suspension performance measure $\tilde{J}$-index*). *Let us define the function $\mathscr{C} : \mathbb{R} \times \mathbb{R} \times \mathbb{R} \to \mathbb{R}$, as follows:*

$$\mathscr{C}(X, \underline{f}, \overline{f}) = \int_{\underline{f}}^{\overline{f}} |X(f)|^2 df \qquad (4.8)$$

where $X(f)$ represents the frequency dependent signal of interest, obtained with Algorithm 1, $\underline{f}$ and $\overline{f}$ are the lower and upper bounds of the frequencies of interest. Then, the comfort and road-holding criteria are respectively defined as:

- *$\tilde{J}_{\ddot{z}}$, Comfort criterion (or $\tilde{J}_c$):*

$$\tilde{J}_{\ddot{z}} = \frac{\mathscr{C}(\tilde{F}_{\ddot{z}}, 0, 20)}{\mathscr{C}(\tilde{F}_{\ddot{z}}^{nom}, 0, 20)} \qquad (4.9)$$

- *$\tilde{J}_{z_{def_t}}$, Road-holding criterion (or $\tilde{J}_{rh}$):*

$$\tilde{J}_{z_{def_t}} = \frac{\mathscr{C}(\tilde{F}_{z_{def_t}}, 0, 30)}{\mathscr{C}(\tilde{F}_{z_{def_t}}^{nom}, 0, 30)} \qquad (4.10)$$

where $\tilde{F}_{\ddot{z}}$ and $\tilde{F}_{z_{def_t}}$ are the variance gains of the considered system obtained using Algorithm 1. Then $\tilde{F}_{\ddot{z}}^{nom}$ and $\tilde{F}_{z_{def_t}}^{nom}$ are the nominal suspension variance gains, i.e. the passive uncontrolled reference suspension gains obtained by Algorithm 1 also.

Definition 4.2 (*Suspension performance measure J-index*). *Let us define the function $\mathscr{C} : \mathbb{R} \times \mathbb{R} \times \mathbb{R} \to \mathbb{R}$, as follows:*

$$\mathscr{C}(X, \underline{t}, \overline{t}) = \int_{\underline{t}}^{\overline{t}} |x(t)|^2 dt \qquad (4.11)$$

where $x(t)$ represents the time dependent signal of interest, obtained with the broad band white noise signal used in the Algorithm 2, $\underline{t}$ and $\overline{t}$ represent the interval limits of interest. Then, the comfort and road-holding criteria are respectively defined as:

- *$J_{\ddot{z}}$, Comfort criterion (or J_c):*

$$J_{\ddot{z}} = \frac{\mathscr{C}(\ddot{z}, \underline{t}, \overline{t})}{\mathscr{C}(\ddot{z}^{nom}, \underline{t}, \overline{t})} \qquad (4.12)$$

- $J_{z_{def_t}}$, *Road-holding criterion (or J_{rh}):*

$$J_{z_{def_t}} = \frac{\mathscr{C}(z_{def_t}, \underline{t}, \overline{t})}{\mathscr{C}(z_{def_t}^{nom}, \underline{t}, \overline{t})} \tag{4.13}$$

where $\ddot{z}$ and z_{def_t} are the time domain response signals obtained using Algorithm 2. Then $\ddot{z}^{nom}$ and $z_{def_t}^{nom}$ are the time domain response signals of the passive uncontrolled reference suspension gains obtained by Algorithm 2 also.

Remark 4.1 (*Performance index interpretation*). *Performance measures given in Definition 4.1 are ratios between the analyzed suspension and the reference suspension configuration (which in our case will always be represented by a passive suspension with a nominal damping of $c = 1500\,Ns/m$). Consequently, an improvement is obtained if the performance measure is below 1. Therefore, if:*

- $J_{\ddot{z}} > 1$ *or $\tilde{J}_{\ddot{z}} > 1$ (resp. $J_{\ddot{z}} < 1$ or $\tilde{J}_{\ddot{z}} < 1$), it means that the analyzed suspension is less (resp. more) comfortable than the reference one.*
- $J_{z_{def_t}} > 1$ *or $\tilde{J}_{z_{def_t}} > 1$ (resp. $J_{z_{def_t}} < 1$ or $\tilde{J}_{z_{def_t}} < 1$), it means that the analyzed suspension provides worse (resp. better) road-holding performances than the reference one.*

4.2.3 Numerical Discussion and Analysis

To illustrate the previously presented frequency domain response computation and the performance index consistency, these metrics are now applied on a numerical example. Let us consider the passive quarter-car model, as in Definition 3.1 with the front motorcycle parameters given in Table 1.2. Then, the following damping values are evaluated:

- Nominal damping value, $c = c_{nom} = 1500\,Ns/m$, which is an average value, representing a nice trade-off between comfort and road-holding performances.
- Soft damping value, $c = c_{min} = 900\,Ns/m$, which is a low damping value, intuitively providing more comfort oriented behavior (indeed, we will see that it is not completely true).
- Stiff damping value, $c = c_{max} = 4300\,Ns/m$, which is a high damping value, intuitively providing more road-holding oriented performances (indeed, we will see here again that it is not completely true).

By applying Algorithm 1, the frequency response diagrams given in Figure 4.3 are obtained.

Then, the following comments holds:

- There exists a natural trade-off between comfort (Figure 4.3 – top) and road-holding (Figure 4.3 – bottom) performances. Indeed, it is notable that:

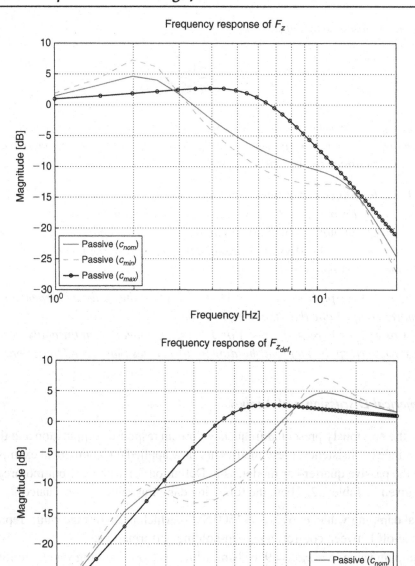

Figure 4.3: Nonlinear frequency responses (FR, obtained from Algorithm 1) of the passive quarter-car model for varying damping values: nominal $c = 1500$ Ns/m (solid line), soft $c = c_{min} = 900$ Ns/m (dashed line) and stiff $c = c_{max} = 4300$ Ns/m (solid rounded line). Comfort oriented diagram $\tilde{F}_z$ (top) and Road-holding oriented diagram $\tilde{F}_{z_{def_t}}$ (bottom).

- No passive suspension configuration can provide both comfort (F_z) and road-holding ($F_{z_{def}}$) optimal characteristics.[1]
- No passive suspension configuration can globally improve filtering for either comfort (F_z) or road-holding ($F_{z_{def}}$) characteristics.

• More specifically, the frequency space may roughly be divided in three zones:
- Low frequencies (in this case $f \in [0; 2]$Hz), where the best filtering is achieved by the stiff damping configuration ($c = c_{max}$).
- Middle frequencies (in this case $f \in [2; 10]$Hz), where the best filtering is achieved by the soft damping configuration ($c = c_{min}$).
- High frequencies (in this case $f \in [10; \infty]$Hz), where the best filtering is achieved by:
  * The stiff damping configuration ($c = c_{max}$), for road-holding objectives (bottom frame).
  * The soft damping configuration ($c = c_{min}$), for comfort objectives (upper frame).

Let us note that, as illustrated in Chapter 5, this frequency space division is not so clear.

From that first frequency analysis, it appears that an optimal control should be able to adjust the damping coefficient according to the road disturbance excitation (see Chapter 5 for more details). This analysis, justifies at a relatively high level, the use of a controlled damper to enhance the vehicle performances. From this analysis, it also appears that there is always a trade-off between comfort and road-holding performances. This trade-off simply means that a passive suspension cannot achieve both comfort and road-holding performances at the same time, but, a controlled one should globally be better for both objectives. In Chapter 5, we will see that controlled dampers cannot achieve optimal comfort and road-holding performances at the same time.

Performance index is now investigated for the three damping configurations simulated on Figure 4.3 (i.e. either $c = c_{nom} = 1500$ Ns/m, $c = c_{min} = 900$ Ns/m and $c = c_{max} = 4300$ Ns/m). Results are plotted on Figure 4.4.

According to the criteria in Definition 4.1, the suspension with nominal damping is selected as the reference (nominal case). In Figure 4.4, both $\tilde{J}_c$ and $\tilde{J}_{rh}$ indexes of the passive nominal damper are equal to 1. Then, according to the results of Figure 4.4, the following remarks hold:

• For a stiff suspension ($c = c_{max}$), the comfort criterion ($\tilde{J}_c$) is higher than one, hence the car is uncomfortable (at least, less than the reference model), while the road-holding criterion ($\tilde{J}_{rh}$) is sightly lower than one, and hence provides better road-holding properties than the passive one.

---

[1] By optimal, one means, a better filtering i.e. a lower bound of the frequency response over all frequencies of interest; see Chapter 5 for more details.

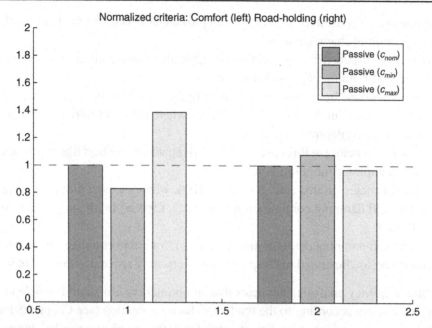

Figure 4.4: Normalized performance criteria comparison for different damping values. Comfort criterion – $\tilde{J}_c$ (left histogram set) and road-holding criterion – $\tilde{J}_{rh}$ (right histogram set).

- Conversely, when a soft suspension ($c = c_{min}$) is considered, the comfort criterion ($\tilde{J}_c$) is lower than one, hence the car is comfortable, while the road-holding criterion ($\tilde{J}_{rh}$) is higher than one, and hence provides worse road-holding performances than the passive one.

Based on the same idea, these criteria can be evaluated for a large number of frozen damping values c, leading to another interesting way to represent the trade-off between comfort and road-holding. On Figure 4.5, the performance indexes $\tilde{J}_c$ and $\tilde{J}_{rh}$ are evaluated for varying damping values (here $c = [100, \ 10,000)$). This figure illustrates the performance trade-off curve of the passive suspension system. Each point of the curve gives the passive quarter-car model performance associated to a damping value; more specifically:

- the abscissa gives the comfort index ($\tilde{J}_c$);
- the ordinate gives the road-holding index ($\tilde{J}_{rh}$);
- while the line intensity changes as a function of the damping value c, (the darker, the lower the damping is).

As defined, the point of coordinate $(1; 1)$ represents the passive nominal case. Then, both passive soft ($c = c_{min}$) and stiff ($c = c_{max}$) corresponding points are plotted on Figure 4.5. The interesting point to notice on this curve is that, as previously indicated by the frequency responses, the increase of the damping coefficient (i.e. $c \to \infty$) will not improve the

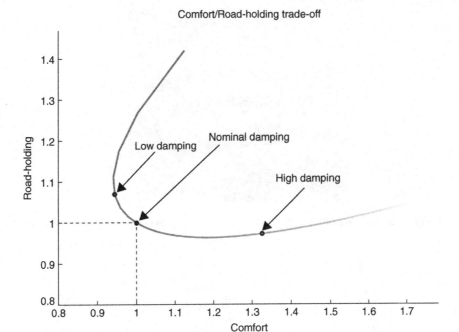

Figure 4.5: Normalized performance criteria trade-off ($\{\tilde{J}_c, \tilde{J}_{rh}\}$) for a passive suspension system, with varying damping value $c \in [100, 10\,000]$ (solid line with varying intensity). Dots indicate the criteria values for three frozen damping values (i.e. $c = c_{min} = 900\,\text{Ns/m}$, $c = c_{nom} = 1500\,\text{Ns/m}$ and $c = c_{max} = 4300\,\text{Ns/m}$).

road-holding ($\tilde{J}_{rh}$), and conversely, the reduction of the damping value (i.e. $c \to 0$) will not lead to a more comfortable vehicle ($\tilde{J}_c$). Therefore there exists an inherent trade-off for uncontrolled suspension. As we will see in Chapter 5, this trade-off is always true, even when damper is perfectly controlled.

Remark 4.2 (*About the trade-off curve*). *In the following chapters, this representation will be re-used to:*

- *evaluate the best performances a controlled semi-active suspension system can achieve (Chapter 5);*
- *evaluate how semi-active strategies can improve passive suspension.*

4.3 Other Time Domain Performance Evaluations

Up to now, only frequency domain analysis has been presented, in order to evaluate the global improvement. Here, time domain experiments are provided. The objective of these time domain experiments is to extend the frequency domain results, by evaluating the

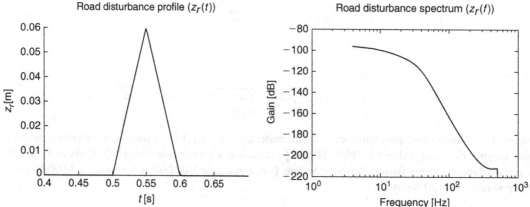

Figure 4.6: Bump road disturbance (top) and its time and frequency representation (bottom left and right respectively).

suspension system subject to realistic road disturbances. The interest of these experiments is that they are more easily understandable, and that they excite many frequencies at the same time, while previous frequency tests excite one single frequency each time.

4.3.1 Bump Test

The bump test refers to a bump on the road which might be viewed as a high frequency excitation. The interest of this time domain experiment is that it is very likely to realize in practice and to interpret. However, this test is only a complement to the previous ones, since it cannot represent all the suspension behaviors. Figure 4.6 illustrates a typical bump test profile and its associated spectrum.

In the following, the previous test bench bump experiment is evaluated on the passive quarter-car model using motorcycle parameters. The experiments are done for two damping configurations:

- Soft damping: $c = c_{min}$.
- Stiff damping: $c = c_{max}$.

According to results given on Figure 4.7, the same comments as on the previous frequency results can be made. Basically, soft damping values provide lower accelerations and an oscillatory behavior, while stiff ones provide high accelerations and less oscillations. This

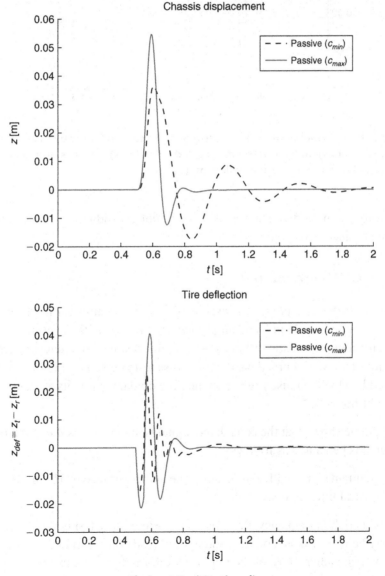

Figure 4.7: (*Continued*)

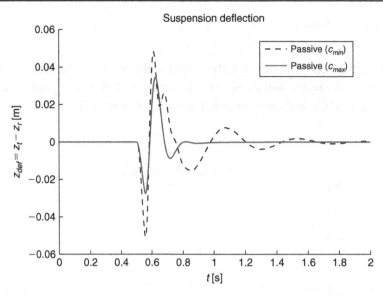

Figure 4.7: Road bump simulation of the passive quarter-car model for two configurations: hard damping (c_{max}, solid lines) and soft damping (c_{min}, dashed lines). Chassis displacement ($z(t)$), tire deflection ($z_{def_t}(t)$) and suspension deflection ($z_{def}(t)$).

experiment is only given for practical issues, but will not give additional information compared to the frequency response.

4.3.2 Broad Band White Noise Test

The white noise disturbance analysis consists in exciting the quarter-car model with a random road profile with limited bandwidth and amplitude, and analyzing the resulting performances. Since it represents a more realistic driving situation, this test is very interesting from the application point of view. The broad band white noise test is given on Figure 4.8. On the left figure, the broad band white noise is given in the time domain, while its associated spectrum is given on the right frame.

The frequency frame shows that the considered signal is attenuated as the frequency increases. Indeed this signal is generated as follows:

- Generate a random signal with zero mean value and a variance of 10 cm.
- Apply an integral filter to this signal.

Since this signal is rich in frequency, the broad band white noise test is very interesting. However, it may be regarded as repetitive and hard to read compared to the frequency response (FR) which gathers most of the important information a control engineer is interested in. As a matter of fact, in this book, this experiment will not always be performed.

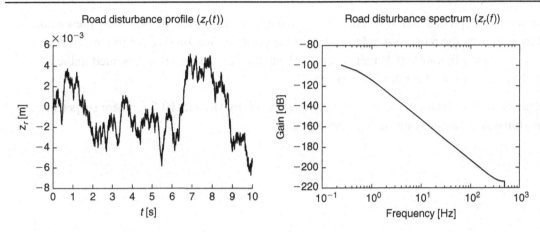

Figure 4.8: Broad band white noise example. Time response (left) and its spectrum (right).

4.4 Conclusions

In this chapter, specific frequency and time domain methodologies to analyze a given suspension systems, are presented. These methodologies include both numerical tools and specific signals to characterize comfort and road-holding performances using the single-quarter vehicle model (defined in Chapter 3). The interest here is to provide the reader with a set of analysis tools to evaluate a given suspension system (e.g. passive configuration) or semi-active control algorithm efficiency with respect to another or even to the optimal one (see Chapter 5). The presented tools are summarized as follows:

- Frequency response computations – Algorithms 1 and 2: which allow us to plot Bode-like diagrams for nonlinear systems. This frequency plot allows us to clearly characterize the suspension system and is much appreciated by both academic and industry communities since it provides a very rich information set over all the frequencies of interest. This experiment provides a very complete picture of the performances the suspension system can achieve.
- Performance criteria computation – Definitions 4.1 and 4.2: which provide a performance index to indicate whether the system is comfort or road-holding oriented. This performance index represents additional information to the frequency response, providing a fast analysis tool to compare semi-active suspension design and control performances.
- Bump test road profile: which allows us to simply understand suspension behavior.
- Noisy broad band road profile: which allows us to evaluate the performance of the suspension on a realistic road disturbance profile, exciting many different frequencies at the same time.

The presented analysis tools will now be used to evaluate different control algorithms (see Chapter 5 to 8 and Appendix B). This tool presentation completes the previous chapters contents (Chapters 2 and 3) where controlled damper and vehicle models were presented. From now on, the reader is ready to analyze the principal semi-active control approaches existing in the literature (Chapter 6) and evaluate the new approaches presented in the following chapters (Chapters 5, 7 and 8).

The reader may also refer to the very complete work by Hrovat (1997), where suspension performance and analysis are well described.

CHAPTER 5

Optimal Strategy for Semi-Active Suspensions and Benchmark

In all theoretical and applicative control fields (as well as in any engineering and mathematical domain), it is very useful and convenient to evaluate what are the best performances a given system, subject to actuator and inner nested limitations, can achieve. This best performance is often referred to as the "optimal bound", and so it will be in this book.

This "optimal bound" is very important when evaluating a control design since it allows the designer to measure how far the controlled system is from the best achievable performance and may help him to adjust the controller parameters according to genuine optimal criteria.

As an illustration, in the time delay control community, a higher theoretical bound of the admissible delay before instability is often considered and compared to control designs (see e.g. Briat, 2008; Niculescu, 2001); similarly, in the automotive braking field, the minimum theoretical braking distance is also often considered as a reference and compared with anti-locking braking systems algorithms (see e.g. Tanelli, 2007). Performance analysis in a more general framework is also treated in Zhou et al. (1996).

In Chapter 4, a methodology to evaluate suspension performances using frequency, time domain and a performance index was proposed. Indeed, this methodology may be viewed as a metric adapted to the suspension problem. According to these performance specifications (or metrics), one aims at evaluating what are the best performances (i.e. the lower bound or "optimal bound") a semi-active suspension system can achieve, considering its technological limitations.

Due to the complexity and the nonlinear phenomena describing the semi-active suspension system, these optimal performances, also called "comfort and road-holding optimal bounds", are not analytically (i.e. not algebraically) calculated but approximated through a numerical optimization approach. In this chapter, two lower bounds (optimal performances or optimal bounds) are studied:

- The comfort optimal performance, called "comfort lower bound".
- The road-holding optimal performance, called "road-holding lower bound".

These lower bounds will then be used throughout the book as benchmark, to evaluate the proposed innovative control algorithms efficiency presented in Chapters 5 to 8 (see also Appendices A and B).

The chapter is organized as follows: Section 5.1 presents the general optimization framework used to compute the optimal bounds. More specifically, the approach used is similar to the Hybrid MPC idea (Bemporad et al., 2003b). Then, Section 5.2 mathematically defines the problem objectives through cost function definitions. System model and logical constraints defining the semi-active suspension problem are given in Section 5.3. Then, the entire problem definition is summarized in Section 5.4. The efficiency of the proposed optimization algorithm is then numerically illustrated on the quarter-car model, using the motorcycle parameters, in Section 5.5. The "optimality"[1] of the results are also qualitatively discussed. Conclusions and discussions of the method are summarized in Section 5.6.

5.1 General Rationale of the Solution

This section aims at describing in detail the general rationale of the proposed optimal strategy seeking. The general objective is first simply described, linking the problem with tools introduced in Chapter 4, then, the optimization idea is described in a global manner, introducing the subproblems to be solved.

5.1.1 Objective and Assumptions

In this chapter, the aim is to describe a numerical method to compute a genuine lower bound of the frequency response of the semi-active suspension problem for comfort or road-holding objectives, using optimization tools. For this purpose, in this chapter (and in this chapter only), the following assumptions will be considered:

1. The road disturbance profile (z_r) is considered as a known variable for a given horizon N. This assumption means that the road profile $[z_r(k), \ldots, z_r(k+N-1)]$ is known in advance.
2. The state variables $(x(k))$ of the system are assumed to be fully and perfectly measured (i.e. no measurement noise).
3. The semi-active quarter-car model (Σ_d) is considered to be perfectly known and perfectly fitting the real system (no system uncertainty).

The control engineer should note that these assumptions are very strong and absolutely not verified when controlling a real semi-active system. However, the reader should remember

---

[1] Note that in this chapter, the word "optimal" is used with a slight abuse of language since no formal optimal proof is given here. However, results given in Section 5.5 will show that even if the optimal bound may not be achieved, it can be assumed to be fairly well approximated.

that the method proposed here does not aim at being an on-line implementable technique but is intended to provide a theoretical lower frequency bound for analysis and benchmarks.

The idea here is neither to apply this technique on a real-time application nor to discuss its robustness properties but to obtain a lower bound for comfort and road-holding objectives. The singularity of the control algorithm proposed here, compared to other (Hybrid, Explicit) MPC control designs found in the literature (involving on-line optimization procedures) such as Canale et al. (2006); Di-Cairano et al. (2007); Giorgetti et al. (2005, 2006); Giua et al. (2004), is that in this study, the road unevenness (z_r) is considered as a known variable (for a given horizon N) and is included in the problem definition (i.e. problem constraints). It is clear that this measure is practically impossible to obtain, but this assumption will greatly help the optimization procedure. Therefore, the authors stress that there is no sense in comparing this work to the MPC controller design provided in the literature; the results here are purely theoretical.

5.1.2 Optimization: General Idea

Since the semi-active quarter-car model presents actuator limitations which may be viewed as variable saturations (i.e. saturation function of the model states, see Chapter 2), the method consists in describing a nonlinear optimization problem with the following elements (indeed the problem will be defined as a mixed-integer optimization one, to take advantage of recent optimization tools):

1. A cost function, representing the performance objectives (see Chapter 4), either comfort or road-holding, to be minimized (described in Section 5.2).
2. A quarter-car model (given in discrete time), which represents the dynamical equality constraints of the optimization problem (described in Section 5.3).
3. A set of logical control inequality constraints guaranteeing the semi-activeness of the actuators (see Chapter 3), i.e. that control signals lie in the damper achievable domain (described in Section 5.3). These constraints are specific for the semi-active application and we will see that they can be described with binary variables in the optimization problem.

Consequently, due to these binary variables, the entire problem results in a Hybrid Dynamical Model (HDM) (Bemporad et al., 2002, 2003a,b; Tondel et al., 2003). According to these elements, on Figure 5.1, the general iterative optimization scheme to compute the optimal comfort and road-holding bounds is shown.

In this figure, the "Optimization Algorithm" takes as inputs the state measure $x(kT_e)$ and the present and future road disturbance $z_r(kT_e, \ldots, (N-1)kT_e)$. Since the system is assumed to be perfectly known, the Optimization Algorithm "only" has to calculate the control input which minimizes the objective function while guaranteeing the constraints. This problem is

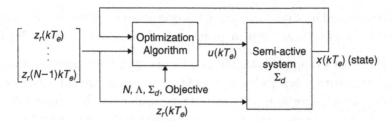

Figure 5.1: **Semi-active suspension optimal performance computation scheme.**

described in a more formal and mathematical way in the following sections, including some numerical remarks the reader should note.

5.2 Cost Function Definitions

According to the performance definitions given in Chapter 4, the objectives of the controlled dampers are to enhance:

1. Comfort performances, by reducing vertical acceleration ($\ddot{z}$) or displacement (z) w.r.t. road disturbances (z_r).
2. Road-holding properties, by reducing the tire deflection ($z_t - z_r$) w.r.t. road disturbances (z_r).

Then, the following cost function definition holds.

Definition 5.1 (*Cost function*). *Let us define the following general quadratic cost function (to be minimized):*

$$J_i\big(N, u(kT_e), x(kT_e), z_r([kTe; (N-1)kT_e])\big) \tag{5.1}$$

where $T_e \in \mathbb{R}^+$ is the sampling time, $k \in \mathbb{N}$ is the sample index, $N \in \mathbb{N}$ is the prediction horizon, $u(kT_e) \in U \subseteq \mathbb{R}$ represents the control input to be computed, $x(kT_e) \in X \subseteq \mathbb{R}$ lies for the system states at time kT_e, $z_r([kT_e; (N-1)T_e])$ is the road disturbance value from present time kT_e until future time $(N-1)kT_e$ and $i = \{c, rh\}$ is the performance objective, defined below. For sake of readability, this function will be denoted as:

$$J_i(N, u, x, z_r) \tag{5.2}$$

Note that in Definition 5.1, the control input $u(kT_e) \in U$. This notation artifact nested the fact that the control input signal should lie in the semi-active set $\mathscr{D}(c_{min}, c_{max}, c^0)$, recalled in Figure 5.2 (left). This specific constraint will be treated in Section 5.3.

According to Definition 5.1, let us now define more specifically the following two cost functions representing either the comfort or the road-holding criteria:

1.   The comfort oriented cost function J_c:

$$J_c(N, u, x, z_r) = \sum_{k=0}^{N-1} \ddot{z}(k)^T \ddot{z}(k) \tag{5.3}$$

which measures the vertical acceleration of the suspended mass M over N samples.

2.   The road-holding oriented cost function J_{rh}:

$$J_{rh}(N, u, x, z_r) = \sum_{k=0}^{N-1} \left(z_t(k) - z_r(k)\right)^T \left(z_t(k) - z_r(k)\right) \tag{5.4}$$

which measures the vertical tire deflection $z_t(k) - z_r(k)$ over N samples.

Note that these two cost functions typically are the function to be minimized in the optimization problem. The reader may also have noted that each objective is related to one of the performance indexes given in Chapter 4.

5.3 Optimization Problem Constraint Definitions

5.3.1 Dynamical Equality Constraints

In order to numerically solve the optimization problem, the dynamical model, represented by equality constraints, should be formulated in discrete-time. This step, theoretically simple, should practically be done carefully. Indeed, in our application case, since the system is of varying dynamics (see Chapter 3), the discretization step should introduce some errors according to the chosen method and sampling-time. As a matter of fact, as suggested in Chapter 3 and Definition 3.7, it is preferable to discretize the semi-active suspension system with a nominal damping value, i.e. discretize $\Sigma_c(c^0)$, as given in Definition 3.7 for non-zero c^0 value to avoid badly damped modes (which would lead to discrete poles close to the limits of the unitary disc). Then, the following discrete-time quarter-car model definition holds:

Definition 5.2 (Discrete-time semi-active quarter-car model – equality constraints). *Let us consider, $\Sigma_c(c^0)$, the continuous-time semi-active quarter-car model described as in Definition 3.7, with $c^0 = \frac{c_{min} + c_{max}}{2}$. By applying the backward Euler method with a sampling time T_e, the resulting discrete-time semi-active quarter-car dynamic is then given as:*

$$\Sigma_d(c^0) : x(k+1) = \left(I_n + A(c^0)\right) T_e x(k) + B T_e \left[z_r(k) \quad u(k)\right]^T \tag{5.5}$$

where $A(c^0) \in \mathbb{R}^{n \times n}$ and $B \in \mathbb{R}^{n \times n_u}$ are the dynamical and input matrices associated with the continuous-time semi-active quarter-car model $\Sigma_c(c^0)$. Then, $x(k)$ represents the discretized state of the semi-active system. For sake of readability, $\Sigma_d(c^0)$ will be denoted as Σ_d.

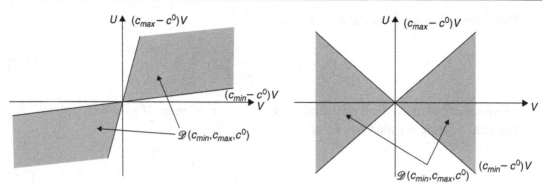

Figure 5.2: Illustration of the domain $\mathscr{D}(c_{min}, c_{max}, c^0)$ modification as a function of c^0. Left: $c^0 = 0$, right: $c^0 = \frac{c_{min}+c_{max}}{2}$

Remark 5.1 (*About c^0 parameter*). *Note that from the numerical point of view, it is crucial to consider a system with nominal damping c^0; if not, the open-loop system will have poles in limit of stability and oscillation, therefore, state variables of the discretized system may explode in simulation (due to numerical errors caused by a stiff model). As a consequence, when the optimization algorithm is used, it may not find any feasible solution. Of course if a nominal damping factor is selected, the damping range has to be modified, i.e., if c^0 nominal factor is considered, the new damping range is no longer $[c_{min}, c_{max}]$ but $[c_{min} - c^0, c_{max} - c^0]$, as illustrated on Figure 5.2).*

Once the nominal damping is set, the main point to be careful with, concerns the sampling time T_e choice. Indeed, it is very important to select a sampling time guaranteeing a faithful frequency representation of the original continuous-time system. For the considered application, a sampling period of $T_e = 1$ ms provides good approximation. Moreover, a nominal damping factor $c^0 = \frac{c_{min}+c_{max}}{2}$ has been selected.

On Figure 5.3, open-loop comparison between the continuous-time and the discrete-time systems, with minimal and maximal dampings is illustrated. Note that concerning the specific semi-active application, the most complex transfer to focus on, is the transfer from the road to the tire deflection since it involves stiff dynamics.

5.3.2 Actuator Inequality Constraints

Now, let us define the inequality constraints of the optimization problem. These (logical) inequality constraints aim at guaranteeing the fact that the control signal lies in the semi-active domain $\mathscr{D}(c_{min}, c_{max}, c^0)$ recalled on Figure 5.2.

Definition 5.3 (*Control constraints – inequality logical constraints*). *Let us define Λ, the set of logic constraints, containing binary variables, ensuring that the control signal (u) lies*

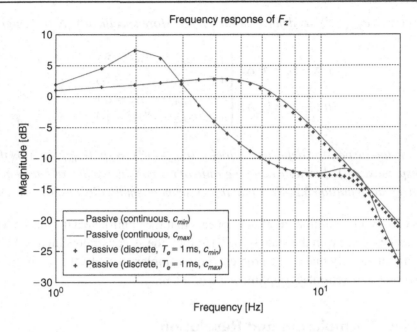

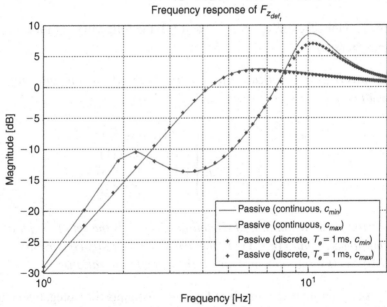

Figure 5.3: Comparison of the continuous and discrete-time (with $T_e = 1$ ms) models frequency responses (Algorithm 1). Top: $\tilde{F}_z$, bottom: $\tilde{F}_{z_{def_t}}$.

in domain $\mathscr{D}(c_{min}, c_{max}, c^0)$, as defined in Chapter 3. More specifically, Λ is then given as:

$$
\begin{aligned}
&if \, \dot{z} - \dot{z}_t \geq 0, \, \Lambda: \begin{cases} u \geq (c_{min} - c^0)(\dot{z} - \dot{z}_t) \\ u \leq (c_{max} - c^0)(\dot{z} - \dot{z}_t) \end{cases} \\
&if \, \dot{z} - \dot{z}_t < 0, \, \Lambda: \begin{cases} u \leq (c_{min} - c^0)(\dot{z} - \dot{z}_t) \\ u \geq (c_{max} - c^0)(\dot{z} - \dot{z}_t) \end{cases}
\end{aligned}
\tag{5.6}
$$

where $\dot{z} - \dot{z}_t$ is the suspension deflection velocity, $(c_{min} - c^0)$ (resp. $(c_{max} - c^0)$) is the new minimal (resp. maximal) allowable damping ratio of the considered discrete-time quarter-car model, as given in Definition 5.2. Note that this Λ set mathematically describes Figure 5.2.

Behind this constraint definition, it clearly appears that the control signal is dependent on the state value, and especially on the state sign. Therefore, the Λ constraints involve binary variables. This singularity makes the problem more complex; namely, the optimization problem becomes a mixed-integer optimization problem.

5.4 Problem Formulation and Resolution

According to the previous Definitions 5.1, 5.2 and 5.3, the following optimization problem, to be solved on-line, stands as:

Definition 5.4 (*Constrained optimization problem*). *Let us define the following constrained finite-time optimal control problem to be solved at each step:*

$$
\begin{aligned}
J_i^*(N, u, x, z_r) = \min \, & J_i(N, u, x, z_r) \\
s.t. \, & \begin{cases} x(0) & = x(k) \\ x(k+1) & = (5.5) \\ \Lambda & = (5.6) \end{cases}
\end{aligned}
\tag{5.7}
$$

where J_i is the criterion to be minimized (Definition 5.1), Λ is the set of inequality logical inequality (Definition 5.3) and $x(k+1)$ are the dynamical equality constraints (Definition 5.2) initialized by $x(0) = x(k)$, the state measure at the given kth iteration.

Since this problem is nonlinear and involves logical constraints (i.e. integer constraints), it is solved using the YALMIP parser (Lofberg, 2004). The solver used in this work is the general optimization solver (GLPK, 2009) (for further details, the reader is invited to refer to the very complete YALMIP wiki[2] and reference therein).

---

[2] http://control.ee.ethz.ch/˜joloef/wiki/pmwiki.php

5.5 Numerical Discussion and Analysis

In this section, the proposed numerical optimization-based optimal semi-active procedure is applied to the semi-active problem (i.e. the quarter-car model with the motorcycle parameters). Then, results are analyzed using the benchmark proposed in Chapter 4. Consequently, the following sections present the frequency results, the performance index evaluation and time simulations showing the efficiency of the method. The frequency responses given hereafter will be considered as the best comfort and road-holding performances, and then used to evaluate the proposed controllers introduced later in this book.

5.5.1 Nonlinear Frequency Response

Applying Algorithm 1, frequency domain results are plotted on Figures 5.4 and 5.5. These figures show the optimal performances (i.e. lower bound) of a controlled semi-active suspension for comfort objective J_c (Figure 5.4) and road-holding objective J_{rh} (Figure 5.5) for varying prediction horizons N. These performances are compared to the passive damper with either c_{min} or c_{max} damping value.

5.5.2 Performance Index

On Figure 5.6, the performance indexes presented in Definition 4.1 are evaluated for varying prediction horizons N, and compared to passive suspensions with stiff, soft and intermediate damping values.

According to the numerical simulations performed, by fixing $N = 15$, the optimal comfort and road-holding performances can reasonably be considered to be reached. Then, by extending the proposed optimal performance analysis, the performance criterion to be minimized is re-defined as follows (with $\alpha \in [0, 1]$):

$$J_\alpha = \alpha J_c(N, u, x, z_r) + (1 - \alpha) J_{rh}(N, u, x, z_r) \tag{5.8}$$

defining a convex combination of the comfort/road-holding performance, allowing us to evaluate the optimal performance trade-off (as for the passive case in Chapter 4). Then, by applying the same procedure as the one given in Definition 5.4 with minimization objective (5.8), and for varying α, the optimal comfort/road-holding performance trade-off can be evaluated. On Figure 5.7, the optimal performance for varying α is plotted and compared with the passive model trade-off and the optimal comfort and road-holding bounds.

5.5.3 Results Analysis and Methodology Discussion

According to the results plotted on Figures 5.4 and 5.5, the following comments can be made:

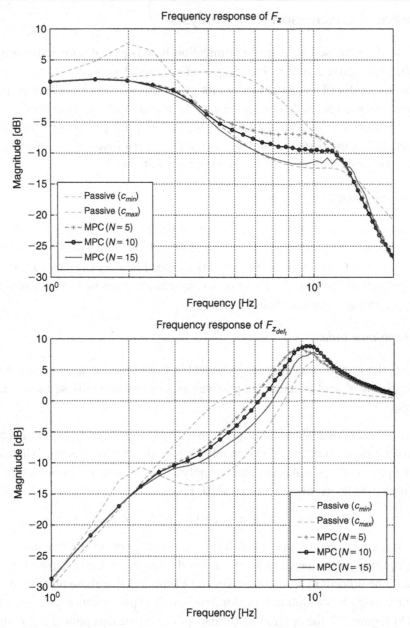

Figure 5.4: Optimal comfort oriented frequency responses of $\tilde{F}_z$ and $\tilde{F}_{z_{def_t}}$ obtained by the optimization algorithm, for varying prediction horizon N, for comfort objective (i.e. cost function $\tilde{J}_c$).

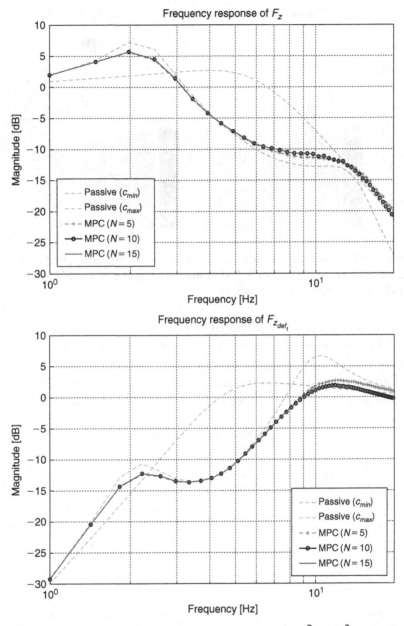

Figure 5.5: Optimal road-holding oriented frequency responses of $\tilde{F}_z$ and $\tilde{F}_{z_{def_t}}$ obtained by the optimization algorithm, for varying prediction horizon N, for road-holding objective (i.e. cost function $\tilde{J}_{rh}$).

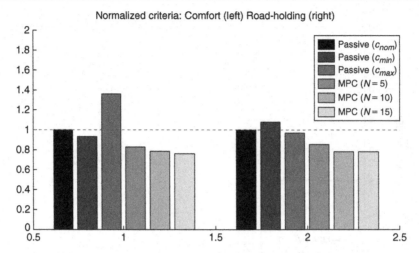

Figure 5.6: Normalized performance criterion comparison for increasing prediction horizon N: comfort criterion — when cost function is $\tilde{J}_c$ (left histogram set) and road-holding criterion — when cost function is $\tilde{J}_{rh}$ (right histogram set).

- By increasing N, the preview and prediction horizon, the lower bound results are improved which is obvious and is completely understandable. Additionally, $N = 15$ seems to be not far from the real optimal performance.
- The trade-off between soft and stiff damping values is quite nicely handled by the optimization algorithm. As a matter of fact, the optimal comfort and handling

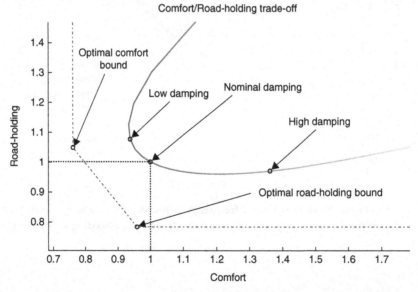

Figure 5.7: Normalized performance criteria trade-off $\{\tilde{J}_c, \tilde{J}_{rh}\}$ for a passive suspension system, with damping value $c \in [c_{min}; c_{max}]$ (solid line with varying intensity) and optimal comfort/ road-holding bounds, with $\alpha \in [0; 1]$ (dash dotted line).

performances are always below the lower passive frequency response of either soft or stiff damping. Additionally, it is notable that, around the invariant points, when a "switch" between high and low damping occurs, the optimal algorithm provides particularly a better performance frequency response. The reader should notice then that since the control law makes the closed-loop nonlinear, these invariant points are modified and the analysis carried out in Chapter 3 does not hold anymore. However, invariant points in $f_1 = 0$ and $f_3 = \sqrt{\frac{k_t}{M+m}}$ remain since they are invariant with respect to the entire suspension force.

- These plots also illustrate the comfort/road-holding trade-off, showing what we already underline in Chapter 4, i.e. the fact that it is impossible to achieve optimal comfort and road-holding performances at the same time.

The results provided on Figure 5.6 illustrate the performance indexes for varying horizon N and compared to passive configuration. On this figure the optimal performance J_c, related to comfort, is plotted for the comfort objective (left histogram set) and J_{rh}, road-holding objective (right histogram set). This figure provides almost the same conclusions as the frequency response diagrams but in a more condensed way. In brief, it illustrates that by controlling the damper in an optimal manner, comfort and road-holding can be greatly enhanced. The increase of the preview horizon N also leads to better performances, which obviously shows saturation over $N = 15$ (equivalent to a preview of 15 ms).

From Figure 5.7, the same conclusions as those of Figure 5.6 can be expressed. More specifically, this figure illustrates the potential of a controlled damper with respect to a passive one evaluated for varying damping coefficients.

Additionally to these comments, concerning the method, it should be borne in mind that results obtained in this section are purely theoretical since in the presented optimal control, the following assumptions have been made:

- The full and perfect knowledge of the system state; which is clearly impossible to achieve in practice since measurements are always noisy and the use of an efficient observer would never guarantee perfect state reconstruction (especially for such systems subject to stochastic disturbances).
- The perfect knowledge of the system parameters and a perfect actuator.
- The perfect knowledge of the road disturbance at time kT_e and its future values until $(N-1)kT_e$ which is impossible on real application.

Moreover, regarding technique validity, the following points should be borne in mind by the reader:

- The optimization algorithm involves binary variables, and results in a highly nonlinear problem. As a consequence, the computed optimal solution is always local and convergence time is unbounded (in comparison e.g. to semi-definite problems where an optimal solution is found in a time that is polynomial in the program description size and of the tolerance error).

- The computed lower bounds are the best achievable performances for this horizon N and sampling time T_e. Actually, the real theoretical lower bound may be theoretically obtained for $N \to \infty$ and $T_e \to 0$ which from a numerical point of view would be very unlikely to be solved. In this study maximal/lower $\{N, T_e\}$ couples were set to $\{15, 1\,\text{ms}\}$. Further tests carried out (not given here) show that the increase of N does not greatly enhance the results (but still increases the computation time).
- Regarding the time computations, these frequency experiments were carried from 1 to 30 Hz, with 60 points, and period number $P = 5$ (see Algorithm 1).

5.5.4 Bump Test

As a complement to the previous study, both optimal comfort and road-holding controller with preview are evaluated on the road bump test presented in Chapter 4 (see Figure 5.8).

5.6 Conclusions

In this chapter, an innovative approach to evaluate the best theoretical performances of a semi-active suspension is presented. Such an approach is based on some (practically unrealistic) assumptions and on an optimization algorithm. Even if the global optimum cannot be theoretically proved (due to the mixed integer nonlinear problem formulation), the results illustrate that the performance tends to the optimal bound. This algorithm has been validated on both frequency and time domain experiments, allowing us to validate on a nonlinear closed-loop system the performance evaluation tools presented in Chapter 4.

The main interest in this work is two-fold: first, it provides a method to analyze and derive the optimal performance that a controlled-damper based suspension system can achieve (which will then be very useful to design and evaluate control algorithms). Second, this chapter illustrates the consistency of the proposed metrics to evaluate semi-active suspension. The results obtained in the present chapter will then play the role of benchmark throughout the book (see also Appendix A).

Additionally to this practical and very useful conclusion, the reader may have noticed that such a procedure can "easily" be applied to other semi-active suspension models. Indeed, such an optimal performance evaluation has been evaluated on a quarter-car model where the control input is simply a linear damping ratio value, but more complex semi-active damper

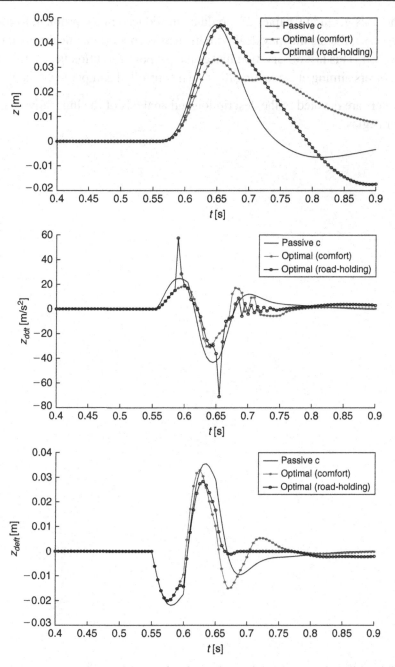

Figure 5.8: Bump test responses of the optimal comfort oriented control (solid small round symbol), optimal road-holding oriented (solid large round symbol) and passive with nominal damping value (solid line). From top to bottom: chassis displacement (z), chassis acceleration ($\ddot{z}$) and tire deflection (z_{def_t}).

models (as the ones given in Chapter 2) should be introduced in the problem formulation. Then in such a case, the results may deal with the new damper dynamical constraints and limitations as well. This last observation makes the proposed solution very interesting for the industrial engineers aiming at evaluating their own controlled damper technology.

The next chapters are devoted to the description and analysis of the innovative semi-active suspension strategies.

CHAPTER 6

Classical Control for Semi-Active Suspension System

The suspension system has been considered in many case studies of control design. Most of the papers have been concerned with active suspensions, since they allow us to obtain greater performances while the control synthesis does not require some dissipativity properties to be handled.

The semi-active suspension control literature is also quite large, and an important number of control strategies exist for such a system. In this chapter, the most important or at least the most developed ones, are recalled together with some "ad hoc" references. This chapter aims at presenting and evaluating some of the usual existing control strategies. The emphasis is mainly put on performance analysis, using the tools described in the previous chapters (see Chapter 4), rather than on an exhaustive and complete description.

The chapter is organized as follows: in Sections 6.1 and 6.2, some usual and simple semi-active suspension strategies, focusing on comfort and road-holding respectively, are presented and evaluated using the frequency response diagrams. Then, in Section 6.3, these strategies are compared using the performance criteria presented in Chapter 4 and briefly discussed. In Section 6.4, some modern semi-active methods are also recalled but, due to their "complexity" (especially for performance tuning), only briefly evaluated. Finally, some conclusions and discussions are given in Section 6.5.

6.1 Comfort Oriented Semi-Active Control Approaches

In this section, the most common comfort oriented semi-active suspension control strategies are presented and evaluated.

6.1.1 Skyhook Control

The principle of this approach is to design an active suspension control so that the chassis is "linked" to the sky in order to reduce the vertical oscillations of the chassis and the axle

DOI: 10.1016/B978-0-08-096678-6.00006-7

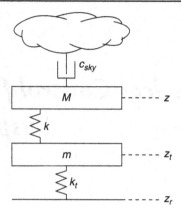

Figure 6.1: Skyhook ideal principle illustration.

independently of each other (Karnopp et al., 1974). Thus a fictitious damper is considered between the sprung mass and the sky frame, as shown in Figure 6.1.

Through the isolation of the sprung mass from the road profile, it allows a reduction of vibration. This desired behavior is then modeled as:

$$\begin{cases} M\ddot{z} = -k(z - z_t) - c_{sky}\dot{z} \\ m\ddot{z}_t = k(z - z_t) - k_t(z_t - z_r) \end{cases} \tag{6.1}$$

where c_{sky} is the damping coefficient of the Skyhook behavior.

Since this is not theoretically possible, this "ideal" system is realized, starting from the model (3.46), using a damper force $c_{sky}\dot{z}$, that allows us to reproduce the Skyhook behavior for the sprung mass (but not for the unsprung mass).

When semi-active dampers are concerned, the active behavior allowing us to obtain the Skyhook phenomenon may be then approached by using semi-active dampers only, as in Emura et al. (1994).

Below, two well-known cases of Skyhook control are described: the discontinuous case, also referred to as the 2-states Skyhook control and the Skyhook linear case.

6.1.1.1 The 2-States Skyhook Control (SH 2-States)

The 2-states Skyhook control is an on/off strategy that switches between high and low damping coefficients in order to achieve body comfort specifications. This control law consists in changing the damping factor c_{in} of the damper (i.e. its fluid viscosity, air resistance, etc.) according to the chassis velocity ($\dot{z}$) and the suspension deflection velocity ($\dot{z}_{def}$) by using a logical rule as:

$$c_{in} = \begin{cases} c_{min} & \text{if } \dot{z}\dot{z}_{def} \leq 0 \\ c_{max} & \text{if } \dot{z}\dot{z}_{def} > 0 \end{cases} \tag{6.2}$$

where c_{min} and c_{max} are the minimal and maximal damping factors achievable by the considered controlled damper, respectively (and usually $c_{max} = c_{sky}$). Then, basically, this control law consists in a switching controller which deactivates the controlled damper when the body speed and suspension deflection speed have opposite signs. The controlled damper technology only needs to have two damping coefficient states.

Many studies have been carried out on the Skyhook control strategy since it represents a simple way to achieve a good comfort requirement (Simon, 2001), as for instance the no-jerk version (Ahmadian et al., 2004). Some extended versions of the Skyhook control have been also developed, such as the adaptive one in Song et al. (2007) to MR dampers or the gain-scheduled one in Hong et al. (2002).

This control strategy presents the advantage of being simple, but it requires two sensors.

6.1.1.2 Skyhook Linear Approximation Damper Control (SH Linear)

An improved version of Skyhook control has been used to handle variable damping, either with discrete damping coefficients, or with a continuously variable damper, as illustrated in Sammier et al. (2003); Sohn et al. (2000).

The linear approximation of the Skyhook control algorithm, adapted to semi-active suspension actuators, consists in changing the damping factor c_{in} according to the chassis velocity ($\dot{z}$) and the suspension deflection (z_{def}) s.t.:

$$c_{in} = \begin{cases} c_{min} & \text{if } \dot{z}\dot{z}_{def} \leq 0 \\ \mathbf{sat}_{c_{in} \in [c_{min}; c_{max}]} \left(\dfrac{\alpha c_{max}\dot{z}_{def} + (1-\alpha)c_{max}\dot{z}}{\dot{z}_{def}} \right) & \text{if } \dot{z}\dot{z}_{def} > 0 \end{cases} \tag{6.3}$$

where c_{min} and c_{max} are the minimal and maximal damping factors achievable by the considered controlled damper respectively. $\alpha \in [0, 1]$ is a tuning parameter that modifies the closed-loop performances. More specifically, when $\alpha = 0$, this control law is equivalent to the two-states Skyhook control.

As the two-states control, the linear approximation consists in a two-modes switching controller which modifies the damping factor according to the body speed and to the suspension deflection speed. The innovation relies on the fact that, according to the second expression (when $\dot{z}\dot{z}_{def} > 0$), such a control provides an infinite number of damping coefficients. As a matter of fact, this control law requires a continuously variable controlled damper (e.g. an MR damper).

Other kinds of dampers with a finite number of damping coefficients can be considered. In the latter case, the force applied by the damper is chosen to be as close as possible to the force required by the controller, as shown in Sammier et al. (2003).

From the computational point of view, this control law also requires two measurements and is simple to implement. Moreover, as seen later, it provides very good performances and may allow for on-line adaptation between soft and hard suspension, thanks to the single coefficient α.

6.1.2 Acceleration Driven Damper Control (ADD)

The ADD control is a semi-active control law described in Savaresi et al. (2004, 2005b), which consists in changing the damping factor c_{in} as:

$$c_{in} = \begin{cases} c_{min} & \text{if } \ddot{z}\dot{z}_{def} \leq 0 \\ c_{max} & \text{if } \ddot{z}\dot{z}_{def} > 0 \end{cases} \tag{6.4}$$

where c_{min} and c_{max} are the minimal and maximal damping factors achievable by the considered controlled damper respectively.

This strategy is shown to be optimal in the sense that it minimizes the vertical body acceleration when no road information is available (therefore, this control law is a comfort oriented one). Since it requires the same number of sensors as the Skyhook 2-states and the linear approximation control law, this control law is simple from the implementation point of view. Note that the control law is very similar to the 2-states approximation of the Skyhook algorithm, with the difference that the switching law depends on body acceleration ($\ddot{z}$), instead of body speed (which is easier to measure on real vehicles). It is worth noting that the ADD design is well adapted to comfort improvement but not to road-holding. Moreover, the "switching dynamic" may influence the closed-loop performances.

6.1.3 Power Driven Damper Control (PDD)

In Morselli and Zanasi (2008), the authors propose a semi-active suspension control strategy using the port Hamiltonian techniques, which provide powerful tools for modeling mechatronics systems with dissipative components. Based on this observation, it is straightforward that this framework fits the semi-active suspension problem, where the actuator is clearly strictly dissipative. The Power Driven Damper (PDD) control approach is described by the equation (6.5):

$$c_{in} = \begin{cases} c_{min} & \text{if } kz_{def}\dot{z}_{def} + c_{min}\dot{z}_{def}^2 \geq 0 \\ c_{max} & \text{if } kz_{def}\dot{z}_{def} + c_{max}\dot{z}_{def}^2 < 0 \\ \dfrac{c_{min} + c_{max}}{2} & \text{if } z_{def} \neq 0 \text{ and } \dot{z}_{def} = 0 \\ -\dfrac{kz_{def}}{\dot{z}_{def}} & \text{otherwise} \end{cases} \tag{6.5}$$

where k is the stiffness of the considered suspension.

In Morselli and Zanasi (2008) (see also Figure 6.2), the authors show that this strategy provides results comparable to those of the ADD control law, while avoiding the chattering effect of the damping control value. Then, the advantages of the PDD control are clearly linked to this non-chattering phenomenon (not illustrated here, but the reader may refer to Morselli and Zanasi (2008)). The additional cost is the need for the knowledge of the spring stiffness k.

6.2 Road-Holding Oriented Semi-Active Control Approaches

Very few studies have been devoted to the possible improvement of road-holding, using a suspension actuator. Indeed the usual main issue in suspension control is ride comfort for which such "actuators" have been considered. In recent years, studies on global chassis control have emphasized that the suspension system may also help in obtaining better road-holding and even handling.

Among the papers dedicated to road-holding characteristics, the Groundhook approach, which is in some sense the dual of the Skyhook one, is described below. It consists in increasing the damping, then reducing the deflection, to reduce the road-tire forces, as illustrated in Valasek and Kortum (2002). As in the Skyhook case, the ideal Groundhook cannot be achieved and needs to be approximated.

Both cases of the 2-states and continuous Groundhook strategies are recalled below. The reader may refer to Koo (2003) and Koo et al. (2004a) where more details are given about this control method (even if this concerns floor vibration applications and not vehicle ones), and where some extended cases, such as displacement and velocity based Groundhook control policies, are presented and tested experimentally.

6.2.1 Groundhook Damper Control (GH 2-States)

In a dual way to the Skyhook case, the 2-states Groundhook control consists in a switching control law depending now on the sign of the product between the suspension deflection

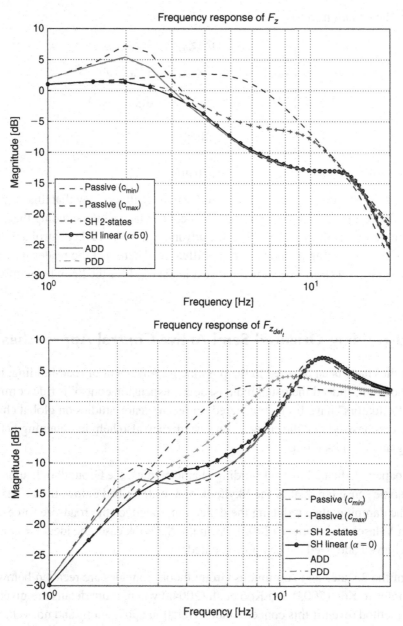

Figure 6.2: Comfort oriented control law frequency responses F_z (top) and $F_{z_{def_t}}$ (bottom).

velocity $\dot{z}_{def}$ and the velocity of the unsprung mass $\dot{z}_t$, as given below:

$$c_{in} = \begin{cases} c_{min} & \text{if } -\dot{z}_t \dot{z}_{def} \leq 0 \\ c_{max} & \text{if } -\dot{z}_t \dot{z}_{def} > 0 \end{cases} \tag{6.6}$$

where c_{min} and c_{max} are the minimal and maximal damping factors achievable by the considered controlled damper respectively.

6.2.2 Groundhook Damper Control (GH Linear)

In this case, the semi-active damper allows us to continuously change the damping coefficient, according to:

$$c_{in} = \begin{cases} c_{min} & \text{if } -\dot{z}_t \dot{z}_{def} \leq 0 \\ \text{sat}_{c_{in} \in [c_{min}; c_{max}]} \left(\dfrac{\alpha c_{max} \dot{z}_{def} + (1-\alpha) c_{max} \dot{z}_t}{\dot{z}_{def}} \right) & \text{if } -\dot{z}_t \dot{z}_{def} > 0 \end{cases} \tag{6.7}$$

where c_{min} and c_{max} are the minimal and maximal damping factors achievable by the considered controlled damper respectively. As in the SH linear, $\alpha \in [0; 1]$ is the tuning parameter to adjust the control law performances.

6.3 Performance Evaluation and Comparison

In this part, the control strategies presented before are compared, using the performance analysis criteria defined in Chapter 4.

6.3.1 Comfort Oriented Strategies

As illustrated on Figures 6.2 and 6.3, the Skyhook strategy greatly improves the comfort criteria. Indeed, it allows us to handle the two suspension phases (compression and expansion) in a way that leads to a better behavior than a simple passive suspension (for the whole frequency range), whatever the damping coefficient is.

Note that the SH 2-states gives almost the same results as the linear Skyhook for $\alpha = 0$. For $\alpha = 1$ the linear strategy corresponds to a classical damper with $c_{max} = c_{sky}$ as damping coefficient. Since it is usually high, it corresponds to a road-holding oriented suspension, as emphasized in Figure 6.3.

Finally, as explained before, the ADD strategy gives the better level of comfort but it deteriorates road-holding compared to the passive suspension. Also note that the PDD control approach provides results very similar to those of the ADD.

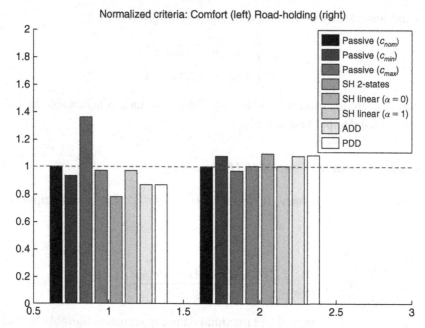

Figure 6.3: Normalized performance criteria comparison for different comfort oriented control strategies: comfort criterion – when cost function is $\tilde{J}_c$ (left histogram set) and road-holding criterion – when cost function is $\tilde{J}_{rh}$ (right histogram set).

6.3.2 Road-Holding Oriented Strategies

The Groundhook strategy is in fact dual with the Skyhook one, and so is the analysis. Hence, according to Figure 6.4, it allows us to obtain better road-holding than a simple passive suspension (for the whole frequency range), whatever the damping coefficient. On the other hand, the comfort is worse than that of both passive cases.

According to the results presented in Figures 6.4 and 6.5, it appears that the Groundhook strategy achieves the best road-holding while it gives the worst comfort over the whole frequency range.

Note that some hybrid approaches, combining Skyhook and Groundhook strategies have been developed, for instance in Flores et al. (2006). However, the switching strategy is not straightforward and needs an accurate tuning to achieve an improvement in both criteria, comfort and road-holding.

6.3.3 About the Trade-Off Comfort Versus Road-Holding

In Figure 6.6, the trade-off between comfort and road-holding is illustrated for all the previous strategies. As presented previously, the ADD and PDD strategies are optimal for comfort, while the GH 2-state is the best for road-holding. SH 2-states give a good comfort (slightly

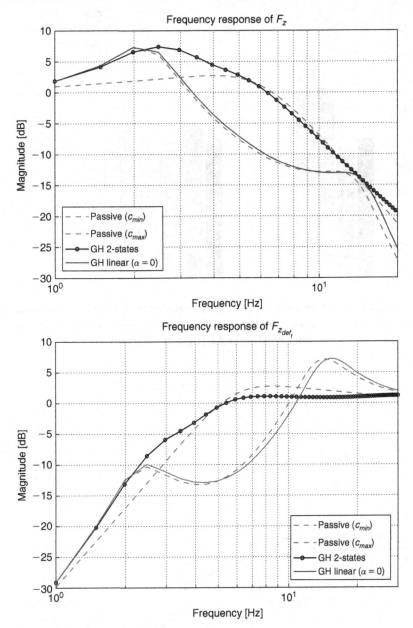

Figure 6.4: Road-holding oriented control law frequency responses F_z (top) and $F_{z_{def_t}}$ (bottom).

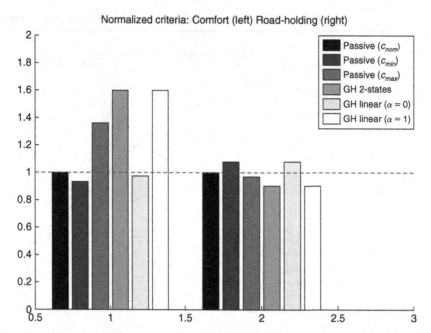

Figure 6.5: Normalized performance criteria comparison for the different road-holding oriented control strategies: comfort criterion – when cost function is $\tilde{J}_c$ (left histogram set) and road-holding criterion – when cost function is $\tilde{J}_{rh}$ (right histogram set).

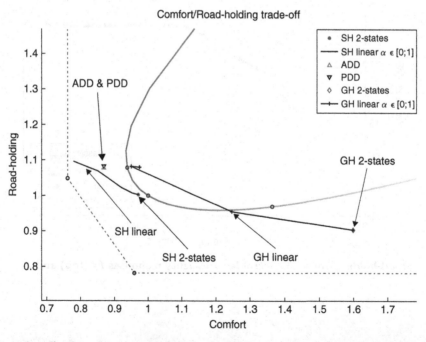

Figure 6.6: Normalized performance criteria trade-off for the presented control algorithms, compared to the passive suspension system, with damping value $c \in [c_{min}; c_{max}]$ (solid line with increasing gray intensity), optimal comfort and road-holding bounds (dash dotted line).

lower than ADD) while the road-holding is close to the passive one. Finally the linear strategies (SH and GH) allow us to achieve various compromises between comfort and road-holding: they may thus allow for on-line variation of that trade-off.

6.4 Modern Semi-Active Control Approaches

In this section, some studies dedicated to semi-active suspension control are presented. Here the emphasis is put on robust or predictive approaches, which are today of great interest in vehicle dynamics.

6.4.1 $\mathcal{H}_\infty$ Clipped Control Approach

Many works have concerned the application of the $\mathcal{H}_\infty$ control approach (sometimes in the framework of LPV systems) for the suspension system. However, most of the results were obtained for active suspensions, as in Zin et al. (2008). When applied to semi-active dampers, the dissipativity constraint of the damper is usually handled using a simple projection (i.e. saturation). While it is not always referred to as "the clipped approach", the latter is very widespread in control strategies for semi-active suspension. Indeed, the aim is very simple: Karnopp (1983) and Margolis (1983) suggested a semi-active controller by passing the optimal active controller through a limiter (saturation). In the control step, the force applied by the semi-active damper is then chosen to be as close to the force required by the controller for a given suspension deflection speed and for the possible range of forces the damper can deliver. This simple strategy has then been applied in many cases, whatever the control method is: optimal, robust, Skyhook, state-feedback... (see also Du et al., 2005b; Rossi and Lucente, 2004; Sename and Dugard, 2003).

In Du et al. (2005a) a static output feedback (clipped) $\mathcal{H}_\infty$ controller is designed for suspension equipped with MR damper. It is shown that it allows us to improve comfort and road-holding as well, compared to a passive suspension. This strategy uses two measured outputs: the suspension deflection and the velocity of the sprung mass.

In Sammier et al. (2003), an $\mathcal{H}_\infty$ methodology is provided so as to be open to any control design engineer in automotive industry. The performance specifications have been expressed as weighting functions on the usual sensitivity functions, which corresponds to a mixed sensitivity problem. The only measured output needed for control is the suspension deflection $(z - z_t)$, which is widely used in the automotive industry because of its low cost and simplicity. The controlled variables are $\{z, z_t, \ddot{z}\}$, allowing us to greatly improve comfort as well as road-holding. The designed $\mathcal{H}_\infty$ controlled is then applied to a semi-active damper available in some PSA Peugeot-Citroën vehicles, for which only a discrete number of damping coefficients can be provided. It is shown that the $\mathcal{H}_\infty$ controlled semi-active suspensions are more efficient than Skyhook ones or passive ones, in terms of comfort and road-holding industrial criteria.

Remark 6.1 (*"Clipped"*). *This control law is now known as "clipped-optimal". The question that arises is: is optimal the clipped-optimal? If not, how far is it from the real optimal one? How would it look like the optimal semi-active one? Clipped approaches lead to unpredictable behaviors and ensure neither closed-loop internal stability nor performances. Active control applied on a semi-active damper results in a "synthesize and try" method, without any performance guarantee (Canale et al., 2006; Giorgetti et al., 2006; Sammier et al., 2003; Tseng and Hedrick, 1994).*

6.4.2 Predictive Approaches

6.4.2.1 Hybrid-Model Predictive Control (Hybrid MPC)

In Giorgetti et al. (2006), the authors introduce an hybrid-model predictive optimal controller (using receding horizon). They solve an off-line optimization process which is a finite horizon optimal regulation problem s.t.:

$$\min_{\xi} J(\xi, x(k)) = \min_{\xi} \left[x^T(N) Q_N x(N) + \sum_{k=1}^{N-1} x^T(k) Q x(k) + y^2(k) \right] \qquad (6.8)$$

subject to,

$$\begin{cases} x(k+1) = Ax(k) + Bu(k) \\ \quad y(k) = Cx(k) + Du(k) \\ \quad 0 \le u(k)\dot{z}_{def}(k) \\ \quad |u(k)| \le \Lambda \end{cases} \qquad (6.9)$$

where Q is a performance index and Q_N is the final weight, as in the optimal control theory. Matrices A, B, C and D in (6.9) define the LTI quarter-car model, Λ is the maximal force allowed by the considered controlled damper and $u(k)\dot{z}_{def}(k) \ge 0$, defines the dissipative constraints. ξ is a vector composed by the sequence of control signals (from 0 to $N-1$) to be applied. Finally, N is the prediction horizon. Giorgetti et al. (2006) show that choosing $N = 1$ leads to performances that are identical to those of the clipped-optimal approach, and by increasing N (e.g. until 40), the performances should be significantly improved. The implemented control law does not involve any optimization procedure since the control algorithm provides a collection of affine gains over a polyhedral partition of the system states x (here $x \in \mathbb{R}^4$) (Borrelli et al., 2003). However, this approach exhibits notable drawbacks:

- The receding horizon control law has to commute between a collection of affine gains. Then, the hybrid controller has to switch between a large number of controllers (function of the prediction horizon). As an illustration for $N = 1$, the control has to switch between 8 regions. For $N = 2$, 62 regions are obtained (which may lead to complex implementation or at least heavy test computation).

- The control law requires full state measurement, which is difficult to obtain in an acceptable way, so that an observer has to be introduced. Consequently, as long as the control law is nonlinear, stability and performances should be checked again.

Hybrid MPC approaches are now increasingly studied, but they still lack of robustness properties and of easy applicability. An illustration of the application of such techniques on semi-active suspensions is presented e.g. in Giorgetti et al. (2005). In this case, the optimization problem is solved off-line and then transformed in a set of explicit controllers which may be implemented.

6.4.2.2 Model Predictive Control Design (MPC)

Thanks to improvements in optimization algorithms, MPC control is being increasingly used in automotive applications. Canale et al. (2006) introduce another MPC semi-active suspension which results in good performances compared to those of the Skyhook and LQ-Clipped approaches. The control algorithm consists in a receding horizon strategy given by the following algorithm steps:

1. Measure $x(k)$, the system state.
2. Solve

$$\min_{\xi} J(\xi, x(k), N_p, N_c) \tag{6.10}$$

   such that semi-active and performance constraints are fulfilled and where N_p is the prediction horizon, N_c, the control horizon and ξ the sequence of control signals.
3. Apply the first element of the solution sequence ξ to the optimization problem as the actual control action.
4. Repeat the whole procedure at time $k + 1$.

Results provided in the paper are interesting and deserve a closer look, but the main drawbacks here, are that:

- It requires an on-line "fast" optimization procedure in the control loop.
- It involves (again) optimal control, full state measurement and a good knowledge of the model parameters is necessary (Giua et al., 2004).

6.4.3 Some Other Approaches

6.4.3.1 Optimal Control

The linear quadratic (state feedback) control has been one of the first methodologies to be applied to suspension control, mainly in the case of active suspensions (Hrovat, 1997), since it allows us to minimize some optimal criteria where the comfort and the road-holding objectives can be considered. Some applications to semi-active suspensions have been proposed using some clipped strategies in Giua et al. (2004); Tseng and Hedrick (1994).

6.4.3.2 Quasi-Linearization Control

The quasi-linearization control approach (Kawabe et al., 1998) consists in a two-step design. The first step is a quasi-linearization, on the basis of the oscillating properties of the suspension system. The second one is a frequency loop-shaping. The first step is aimed at attenuating the oscillation in the lower-frequency region, and the second step at attenuating the oscillation in the higher-frequency region.

6.4.3.3 CRONE Suspension

The CRONE control-system design is a frequency-domain approach for the robust control of uncertain (or perturbed) plants under the common unity-feedback configuration. The open-loop transfer function is defined using integro-differentiation with non-integer (or fractional) order. The required robustness is that of both stability margins and performance, and particularly the robustness of the peak value (called resonant peak) of the complementary sensitivity function.

Oustaloup et al. (1996) proposed a design method which can be applied to passive, semi-active and active suspension systems on the basis of classical control theory. With this method, a quarter-car model in which the vehicle suspension system has been simplified by extracting only its bouncing motion is used.

6.5 Conclusions

This chapter has presented some of the control approaches that have been applied to semi-active suspension systems. Even if it is not complete (due to the large literature in this area), the aim was to emphasize that few strategies are generic and optimal enough to be applied to any kind of semi-active suspension. Indeed the clipped methodology, widely used in the literature, cannot ensure the performances and robustness of the controller since the dissipativity constraint is not accounted for in the design. On the other hand, while the predictive control approaches allow us to take into account the actuator constraints, they are currently limited since the required measurements do not cope with the industrial costs and reliability objectives.

The following chapters are devoted to control strategies where the passivity constraint is taken into account, while being well adapted to the required performances of suspension systems, and providing systematic procedures for control synthesis.

CHAPTER 7

Mixed SH-ADD Semi-Active Control

In this chapter two new control strategies for semi-active suspensions are presented: the Mixed SH-ADD and the 1-Sensor-Mix, first introduced in Savaresi and Spelta (2007) and Savaresi and Spelta (2009). It was proven that these rules are able to perform optimally on the entire range of frequencies (for comfort purposes), with the apparatus and the computational effort comparable with traditional suspension strategies, like SH and ADD. In particular the 1-Sensor-Mix ensures even a sensor reduction, with a negligible loss of performance. A complete numerical analysis on a quarter-vehicle model is herein provided.

This chapter is outlined as follows. In Sections 7.1 and 7.2 the Mixed SH-ADD and the 1-Sensor-Mix rationale are described. The range selector, that is the key idea of these rationales, is described and discussed in Section 7.3. In Section 7.4 a full numerical analysis on the quarter-car model is presented.

7.1 Mixed Skyhook-ADD: The Algorithm

In the previous chapter it has been shown that two classical semi-active control strategies, the SH and the ADD, present complementary behaviors in terms of performance. As it appears in the frequency-domain analysis, the SH provides the best performance at low frequency (around the the body resonance), and the ADD ensures optimality at mid and high frequency (beyond the body resonance). The problem is to develop a simple control strategy capable of mixing the best behavior of SH and ADD, without increasing either the computational effort or the hardware complexity (in terms of required sensors). On this basis, an optimal control strategy is proposed: the mixed SH-ADD. The key idea exploits a very simple but effective frequency range selector, which is able to distinguish the instantaneous dynamical behavior of the suspension; in the case of low frequency dynamics the SH is selected, while the ADD is selected otherwise. The resulting control law is incredibly simple and requires the same apparatus as SH.

Proposition 7.1 (*The rationale of the mixed SH-ADD control*). *According to the quarter-car system description in Definition 3.6, given a two-state damper as defined in*

Chapter 3, the mixed SH-ADD algorithm proposed is given as follows:

$$c_{in} = \begin{cases} c_{max} & if \left[(\ddot{z}^2 - \alpha^2 \dot{z}^2) \leq 0 \ \& \ \dot{z}\dot{z}_{def} > 0 \right] OR \left[(\ddot{z}^2 - \alpha^2 \dot{z}^2) > 0 \ \& \ \dot{z}\dot{z}_{def} > 0 \right] \\ c_{min} & otherwise \end{cases} \qquad (7.1)$$

where $\{c_{min}, c_{max}\} \in \mathbb{R}^+$ are the minimum and maximum damping values, and $\alpha \in \mathbb{R}^+$.

The proposed rationale is extremely simple, since – similarly to SH and ADD – it is based on a simple static rule, which makes use of $\ddot{z}$, $\dot{z}$ and $\dot{z}_{def}$ only. The acceleration $\ddot{z}$ may be measured by an accelerometer body side; the velocity signal $\dot{z}$ can be obtained by integrating $\ddot{z}$. The stroke velocity $\dot{z}_{def}$ may be computed by numerical derivation of the stroke signal measured by a linear potentiometer (alternatively, by a numerical integration of two accelerometers placed at body-side and wheel-side).

It is interesting to observe that (7.1) selects, at the end of every sampling interval, the SH or the ADD rule, according to the current value of $(\ddot{z}^2 - \alpha^2 \dot{z}^2)$:

- if $(\ddot{z}^2 - \alpha^2 \dot{z}^2) > 0$, the ADD strategy is selected;
- otherwise the SH strategy is used.

The amount $(\ddot{z}^2 - \alpha^2 \dot{z}^2)$ hence can be considered as a simple "frequency-range selector". The parameter α represents the frequency limit between the low and the high frequency ranges, and it is the only tuning knob of the control strategy (7.1). Specifically the value of α is set at the crossover frequency (in rad/s) between SH and ADD. Hence, for a standard motorcycle suspension, it has to be selected around 19 rad/s (3 Hz).

In Figure 7.1 the frequency response of the SH-ADD algorithm is shown. Then, on Figures 7.2 and 7.3, the mixed SH-ADD algorithm is compared with the other comfort oriented control strategies (see also Chapter 6). It is possible to conclude that the performances of this algorithm are almost optimal; it is able to provide an almost-perfect mix of the best performance of SH and ADD. As such, notice that its performances are very close to the lower bound. In fact, it is hard to find another control algorithm which performs significantly better, in terms of body-acceleration variance.

7.2 The 1-Sensor-Mix Algorithm

The idea of the frequency-range selector is exploited for the definition of another algorithm: the 1-Sensor-Mix. This rationale is based upon the complementary performances of soft and hard passive suspensions. As discussed in previous chapters, a hard suspension is able to damp optimally the body resonance, but with bad filtering at high frequency. On the contrary,

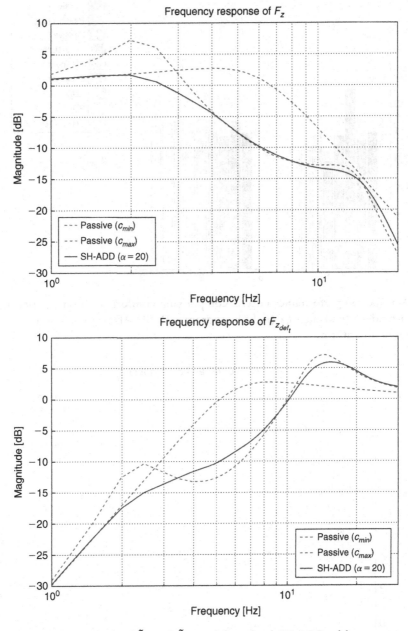

Figure 7.1: Frequency responses of $\tilde{F}_z$ and $\tilde{F}_{z_{def_t}}$ of the mixed SH-ADD with respect to the passive car (with minimal and maximal damping).

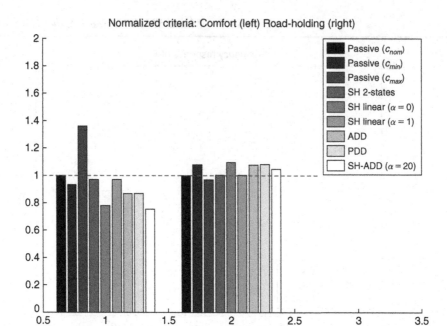

Figure 7.2: Normalized performance criteria comparison: comfort criterion – J_c (left histogram set) and road-holding criterion – J_{rh} (right histogram set). SH-ADD comparison with respect to comfort oriented algorithms.

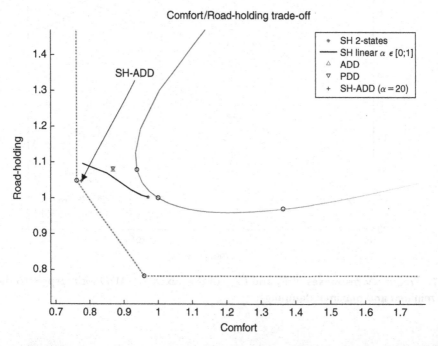

Figure 7.3: Normalized performance criteria trade-off for the presented comfort oriented control algorithms and Mixed SH-ADD, compared to the passive suspension system, with damping value $c \in [c_{min}; c_{max}]$ (solid line with varying intensity), optimal comfort and road-holding bounds (dash dotted line).

a soft suspension ensures the best filtering but with the drawback of a poorly damped body resonance. At the damping invariant point (just beyond the body resonance) the suspension system provides the same performance independently from the damping coefficient (see Chapter 3 – Invariant points, for discussion). This complementarity is similar to the "semi-active complementarity" between the SH and the ADD. However the main difference is that at mid frequency the passive performance is far from the one achievable by the best possible semi-active suspension. Further, out of their range of optimality the semi-active strategies SH and ADD guarantee a better behavior than hard damping and soft damping (see Chapter 5).

Proposition 7.2 (*The rationale of 1-Sensor-Mix SH-ADD control*). *According to the quarter-car system description in Definition 3.6 and given a two-state damper as defined in Chapter 3, the mixing algorithm proposed is given as follows:*

$$c_{in} = \begin{cases} c_{min} & \text{if } (\ddot{z}^2 - \alpha^2 \dot{z}^2) \geq 0 \\ c_{max} & \text{if } (\ddot{z}^2 - \alpha^2 \dot{z}^2) < 0 \end{cases} \tag{7.2}$$

where $\{c_{min}, c_{max}\} \in \mathbb{R}^+$ are the minimum and maximum damping values and $\alpha \in \mathbb{R}^+$.

The proposed control law is extremely simple, since it is based on a simple static rule, which makes use of $\ddot{z}$ and of $\dot{z}$. Note that the acceleration signal $\ddot{z}$ may be acquired by an accelerometer body side, while the velocity signal $\dot{z}$ can be obtained by a numerical integration of $\ddot{z}$. Therefore the use of a single sensor is required.

The rule (7.2) selects, at the end of every sampling interval, the hard damping or the soft damping condition, according to the current value of $(\ddot{z}^2 - \alpha^2 \dot{z}^2)$:

- if $(\ddot{z}^2 - \alpha^2 \dot{z}^2) \geq 0$, the soft damping condition is selected;
- otherwise the hard damping condition is used.

Similarly to the mixed SH-ADD rationale, the amount $(\ddot{z}^2 - \alpha^2 \dot{z}^2)$ can be considered as a simple "frequency-range selector", where the parameter α represents the limit between the ranges of low and high frequency. In this design, the value of α is set at the damping invariant point. Hence, for a standard motorcycle suspension it has to be selected around 20 rad/s (3.2 Hz).

In Figures 7.4, 7.5 and 7.6, the 1-sensor-Mix algorithm is compared with the passive soft, hard suspensions and other comfort oriented semi-active suspension strategies previously presented (see also Chapter 6). Some conclusions may be drawn:

- The performances of this algorithm are quasi-optimal: it is able to provide a quasi-perfect mix of the best performance of hard damping and soft damping. As such, notice that its performances are close to the lower bound. It is outperformed by the mixed SH-ADD only.
- Notice that the 1-Sensor-Mix pays little in terms of performance with respect to the mixed SH-ADD, above all, around both the body and wheel resonances. This is not surprising

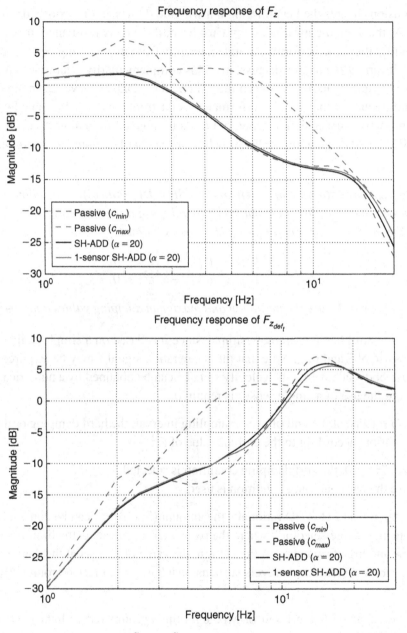

Figure 7.4: Frequency responses of $\tilde{F}_z$ and $\tilde{F}_{z_{def_t}}$ of the mixed 1-sensor SH-ADD with respect to the passive car (with minimal and maximal damping).

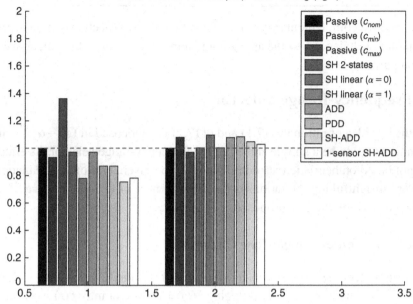

Figure 7.5: Normalized performance criterion comparison: comfort criteria – J_c (left histogram set) and road-holding criterion – J_{rh} (right histogram set). SH-ADD 1-sensor comparison with respect to comfort oriented algorithms.

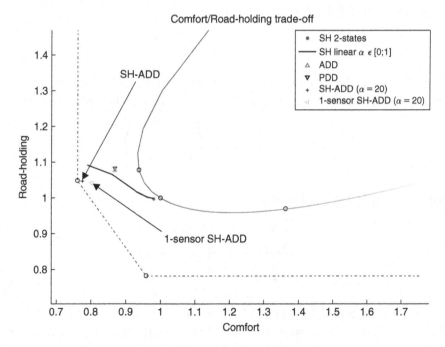

Figure 7.6: Normalized performance criteria trade-off for the presented comfort oriented control algorithms and 1-sensor mixed SH-ADD, compared to the passive suspension system, with damping value $c \in [c_{min}; c_{max}]$ (solid line with varying intensity), optimal comfort and road-holding bounds (dash dotted line).

since the passive complementarity is less clean than the semi-active complementarity. The small loss is balanced by the appealing feature of a sensor reduction, above all for industrial purposes.

7.3 The Frequency-Range Selector

Notice that the key idea of rationales (7.1) and of (7.2) is condensed in $(\ddot{z}^2 - \alpha^2 \dot{z}^2)$; in practice, this function turns out to be a simple and effective frequency-range selector. Its effectiveness however is not based on heuristic reasoning, but it can be given a rational explanation. Two simple but very insightful interpretations of the effectiveness of $(\ddot{z}^2 - \alpha^2 \dot{z}^2)$ as a frequency-range selector are now proposed.

7.3.1 First Interpretation: Single-Tone Disturbance

Consider the single-tone periodic signal $\dot{z}(t) = \dot{A}\sin(\omega t)$. If the function $f(t)$ is defined as $f(t) = (\ddot{z}^2 - \alpha^2 \dot{z}^2)$, where $\alpha \in \mathbb{R}^+$, by plugging $\dot{z}(t) = \dot{A}\sin(\omega t)$ into $f(t)$ we obtain:

$$f(t) = A^2 \omega^2 - A^2 \sin^2(\omega t)(\alpha^2 + \omega^2) \qquad (7.3)$$

Consider now the problem of studying the positivity of $f(t)$ over one period of this function (the period being $T(\omega) = \frac{\pi}{\omega}$). It is easy to see that:

$$f(t) > 0 \Leftrightarrow \sin^2(\omega t) < \frac{\omega^2}{\alpha^2 + \omega^2} \qquad (7.4)$$

If we call $\{D_+(\omega) = t : f(t) > 0, 0 \le t \le T\}$ the domain where $f(t) > 0$, it is easy to see that the measure of this "positivity-domain", say $|D_+(\omega)|$, is given by (see also Figure 7.7):

$$|D_+(\omega)| = \frac{2T}{\pi} \arcsin\left(\sqrt{\frac{\omega^2}{\alpha^2 + \omega^2}}\right) \qquad (7.5)$$

Hence, according to (7.5), the following

$$
\begin{aligned}
\frac{|D_+(\omega)|}{T} &\to 0 \quad \text{if } \omega << \alpha \\
\frac{|D_+(\omega)|}{T} &\to 1 \quad \text{if } \omega >> \alpha \\
\frac{|D_+(\omega)|}{T} &\to \tfrac{1}{2} \quad \text{if } \omega = \alpha
\end{aligned}
\qquad (7.6)
$$

This means that, over a period T, $f(t) > 0$ for more than $T/2$ if $\omega > \alpha$; $f(t) < 0$ for more than $T/2$ if $\omega > \alpha$. Hence $f(t)$, can be considered as a simple frequency-selector, centered around the frequency $\omega > \alpha$: when $f(t) > 0$ we can assume that $\omega > \alpha$; otherwise $\omega < \alpha$. Since the function $\frac{|D_+(\omega)|}{T}$ rapidly saturates towards 0 or 1, when $\omega \ne \alpha$ (see Figure 7.8), a good frequency-selection quality is guaranteed.

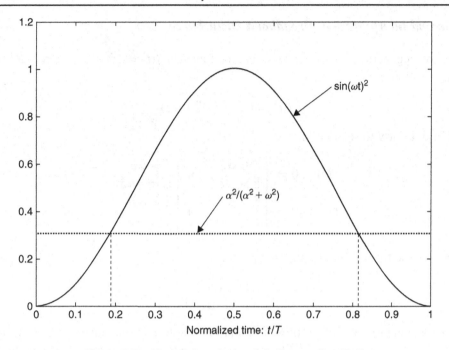

Figure 7.7: Pictorial analysis of the inequality (7.4).

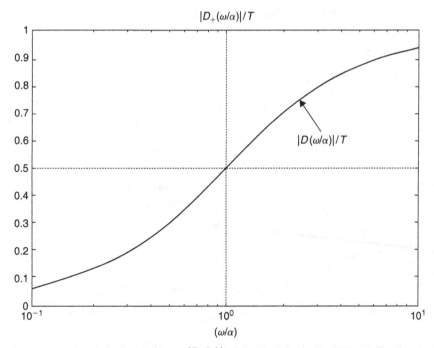

Figure 7.8: Function $\frac{|D_+(\omega)|}{T}$ (in normalized frequency).

7.3.2 Second Interpretation: Broadband Disturbance

Consider now a generic signal $\dot{z}(t)$. Note that the function $f(t) = (\ddot{z}^2 - \alpha^2\dot{z}^2)$, $\alpha \in R^+$, can be rewritten as follows:

$$f(t) = (\ddot{z} - \alpha\dot{z})(\ddot{z} + \alpha\dot{z}) \tag{7.7}$$

From (7.7), it is easy to see that,

$$f(t) < 0 \Rightarrow \begin{cases} \ddot{z} < \alpha\dot{z} \\ \ddot{z} > -\alpha\dot{z} \end{cases} \text{ or } \begin{cases} \ddot{z} > \alpha\dot{z} \\ \ddot{z} < -\alpha\dot{z} \end{cases}$$

$$f(t) > 0 \Rightarrow \begin{cases} \ddot{z} > \alpha\dot{z} \\ \ddot{z} > -\alpha\dot{z} \end{cases} \text{ or } \begin{cases} \ddot{z} < \alpha\dot{z} \\ \ddot{z} < -\alpha\dot{z} \end{cases} \tag{7.8}$$

Consider now the following differential equations:

$$\begin{cases} \ddot{z}(t) = \alpha\dot{z}(t) \\ \ddot{z}(t) = -\alpha\dot{z}(t) \end{cases} \tag{7.9}$$

Note that the equations in (7.9) represent a first-order autonomous linear dynamic system. Since $\alpha \in \mathbb{R}^+$, $\ddot{z}(t) = \alpha\dot{z}(t)$ is unstable, and $\ddot{z}(t) = -\alpha\dot{z}(t)$ is asymptotically stable; note that the position of the pole of this system is at α [rad/s] (which can be considered the cut-off frequency of this dynamic system). The evolution in the time-domain of this linear system, starting from a positive initial condition, is displayed in Figure 7.9.

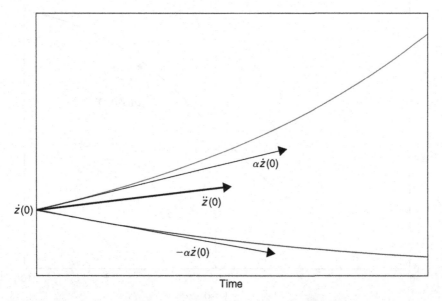

Figure 7.9: Example of evolution of the autonomous systems $\ddot{z}(t) = \alpha\dot{z}(t)$ and $\ddot{z}(t) = -\alpha\dot{z}(t)$ (starting from $\dot{z}(0) > 0$).

Assume that the initial condition $\dot{z}(0) > 0$ (the same reasoning can be made also in the case $\dot{z}(0) < 0$). In this case, note that (7.8) can be simplified as:

$$f(t) < 0 \Rightarrow -\alpha\dot{z}(t) < \ddot{z}(t) < \alpha\dot{z}(t) \tag{7.10}$$

The expression (7.10) can be easily interpreted: note that when $f(t) < 0$ the system evolves with dynamics slower than the cut-off frequency of (7.9); when $f(t) > 0$ the system evolves with dynamics faster than the cut-off frequency of (7.9). Hence, $f(t)$ can be regarded as a simple frequency-separator, even for the case of generic signals.

7.3.3 Sensitivity of Mixed Strategies with Respect to α

As already pointed out, α separates the high and low frequencies domains. It is the only tuning knob of both mixed SH-ADD and 1-Sensor-Mixed.

To the mixed SH-ADD α represents the crossover frequency between SH and ADD (around 3 Hz). To the 1-Sensor-Mixed α represents the crossover frequency between soft damping and hard damping, namely the damping invariant point (around 3.2 Hz).

In Figure 7.10 the performance sensitivity to parameter α is depicted. For the sake of conciseness the analysis focuses on the mixed SH-ADD, but it can be performed equally for the 1-Sensor-Mix. The results confirm the interpretation about α and some conclusions can be drawn:

- If the decision frequency (α) increases, the domain of low frequency becomes wider. So the resulting mixed strategies privilege the low frequency optimal condition. Roughly

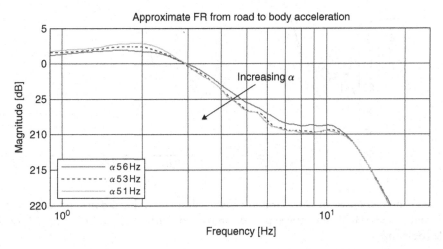

Figure 7.10: Sensitivity to the parameter α of the mixed SH-ADD performances.

speaking SH is preferred by the mixed SH-ADD (the hard damped suspension is preferred by 1-Sensor-Mix). The resulting performances achieve a better damping of the body resonance but a loss of filtering capability at high frequency.

- Overall the sensitivity to α is not such a critical issue. The optimality flavor is maintained. The parameter can be tuned in order to prefer either damping or filtering performances.

7.4 Numerical Time-Domain Simulations

In the previous section, the semi-active control algorithms have been evaluated using the frequency response tool. This analysis tool provides a very clean and condensed picture of the performance of the algorithms. It is interesting, however, to complement the variance gain analysis with time-domain analysis.

7.4.1 Pure Tone Signal

On Figure 7.11 and Figure 7.12, the time-domain responses of the body acceleration when the road profile is a pure-tone signal are displayed. In particular, three single-tone disturbances have been selected: 2.1 Hz (corresponding to the body resonance frequency), 4 Hz

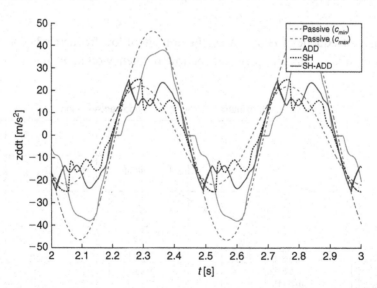

Figure 7.11: Time responses of soft damping suspension (c_{min}), hard damping suspension (c_{max}), SH, ADD, and mixed-SH-ADD to three pure-tone road disturbances: 2.1 Hz (top), 4 Hz (middle) and 12 Hz (bottom).

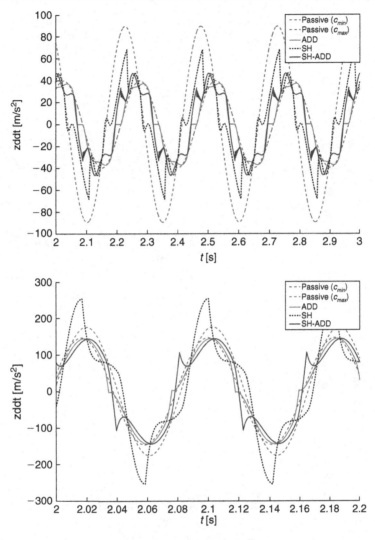

Figure 7.11: (*Continued*)

(representing the intermediate frequency-range) and 12 Hz (representing the high-frequency domain). These results essentially confirm the conclusions drawn from the frequency-domain analysis.

More specifically, by examining Figure 7.11 and Figure 7.12 the following observations may be made:

•   When a semi-active control algorithm is used, the time-domain behavior of the acceleration reveals a very strong nonlinearity of the closed-loop control systems.

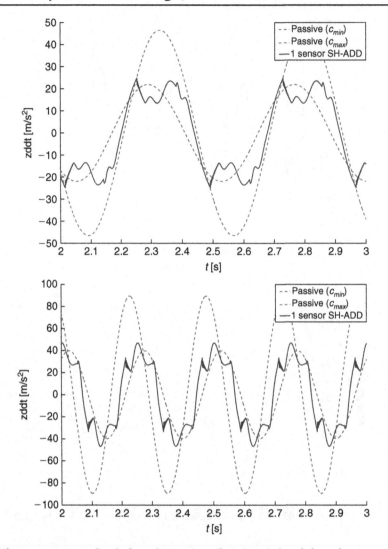

Figure 7.12: Time responses of soft damping suspension (c_{min}), hard damping suspension (c_{max}) and 1-Sensor-Mixed (1SM) to three pure-tone road disturbances: 2.1 Hz (top), 4 Hz (middle) and 12 Hz (bottom).

This confirms that higher harmonics must be taken into account; henceforth – as already stated in Section 7.1 – the variance gain tool is more appropriate than the more classical describing function.

• The SH algorithm and hard damping show superior performance (as expected) at the body resonance frequency. It is clear how SH outperforms ADD at that frequency. Notice, however, that the mixed-SH-ADD and 1-Sensor-Mix algorithms are capable of slightly outperforming SH and hard damping respectively, even at that frequency. It is interesting

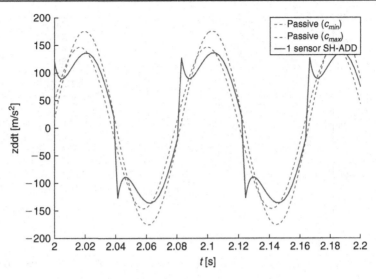

Figure 7.12: (*Continued*)

to notice that both ADD and the mixed algorithms at this frequency show a fast-switching behavior. This chattering phenomenon however is not particularly harmful, since it does not affect the acceleration peaks or the acceleration energy.

- At intermediate and high frequencies the superior performances of ADD and soft damping are confirmed; it is also interesting to notice that at high frequency (12 Hz) the behavior of the closed-loop algorithms tends to become more linear (the higher-harmonics contribution is significantly lower).

7.4.2 Bump Test

The second time-domain analysis refers to the classical response to a triangular bump in the road profile (see analysis in Chapter 4). This kind of excitation is very realistic since it represents a usual 6 cm road bump negotiated at a speed of 30 km/h.

On Figure 7.13 and Figure 7.14 the response of the body acceleration and of the suspension stroke are displayed. In this case, the conclusions which can be drawn are consistent with the previous frequency-domain analysis:

- The SH algorithm and hard suspension provide very good damping, but show the worst acceleration peaks.
- ADD and soft suspension have little acceleration peaks, but the low-frequency damping is very similar to that of the open-loop system.

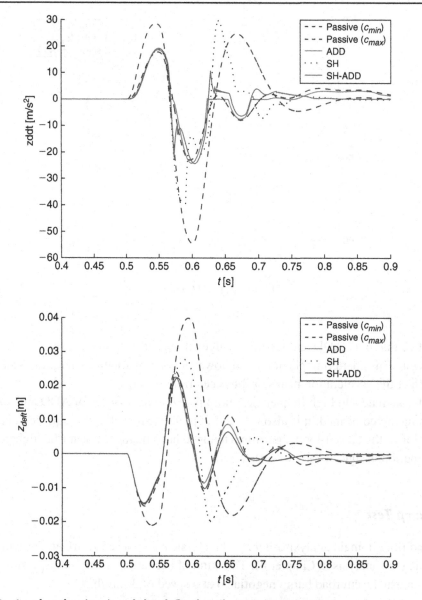

Figure 7.13: Acceleration (top) and tire deflection (bottom) responses to a triangle bump on the road profile: passive soft damping (c_{min}), hard damping (c_{max}), SH, ADD and mixed SH-ADD.

- The mixed SH-ADD algorithm has the same ADD acceleration peaks, but a more efficient damping at low frequencies.
- The 1-Sensor-Mix has the same ADD and soft suspension peaks, but a more efficient damping at low frequency.

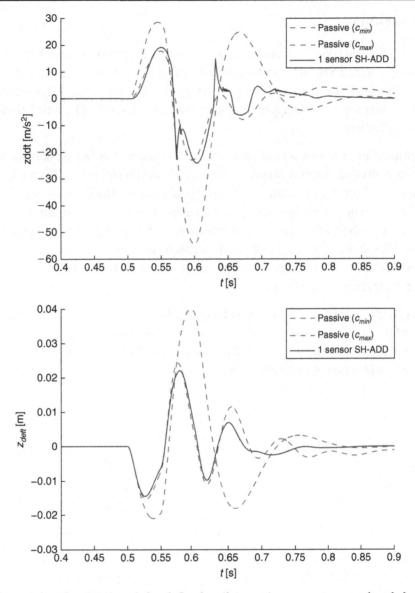

Figure 7.14: Acceleration (top) and tire deflection (bottom) responses to a triangle bump on the road profile: passive soft damping (c_{min}), hard damping (c_{max}) and 1-Sensor-Mixed.

- From the suspension stroke time history, it is clear how in the first part of the transient the mixed SH-ADD shows the behavior of the ADD, and in the second part (after 5.2 s) it shows the behavior of the SH. It is a good example of how a semi-active suspension can adapt its behavior accordingly to the road excitation.

7.5 Conclusions

In this chapter two new control algorithms for semi-active suspensions have been proposed and analyzed: the mixed SH-ADD and the Mix-1-sensor algorithm. The mixed SH-ADD control algorithm provides quasi-optimal performance. The 1-Sensor-Mix has the major feature of providing a very close approximation of the mixed SH-ADD control algorithm, with a sensor reduction.

These algorithms have been presented, and their performances have been analyzed in detail and compared with other comfort oriented semi-active control strategies. The analysis has confirmed the quasi-optimality of the 1-Sensor-Mix algorithm. Thanks to its simplicity and to the fact that it only requires a single sensor (an accelerometer), this algorithm seems to provide the best compromise between costs and performance. Its 1-sensor setting also guarantees superior reliability characteristics. Finally, it is worth observing that the rationale used here for the 1-Sensor-Mix algorithm can be – in principle – extended also to mixed (comfort-handling) control objectives.

In Appendix B, this control law is developed and applied on a real motorcycle illustrating the consistency of the approach. Additionally, in this appendix, some experimental points and practical elements are explained to provide the reader with the background for further implementation and experimental validations.

CHAPTER 8

Robust $\mathcal{H}_\infty$ "LPV Semi-Active" Control

This chapter, which extends results presented by Poussot-Vassal et al. (2007, 2008c), is concerned with the presentation of a robust semi-active suspension control design using a linear parameter varying (LPV) approach. More specifically, the proposed semi-active suspension control strategy is designed so that it minimizes the $\mathcal{H}_\infty$ performance criteria while guaranteeing the limitations of the semi-active damper (i.e. dissipative constraint and force limitations) through a specific parameter dependent structure and a scheduling strategy design. From now on, this controller will be denoted as "LPV semi-active" controller.

This approach exhibits some interesting properties and advantages compared to already existing methods, such as flexibility w.r.t. performance definitions and w.r.t. type of sensors, robustness properties, etc.

The chapter is organized as follows: Section 8.1 recalls the synthesis and the considered controlled damper models, then, the proposed generalized plant used for synthesis and the scheduling strategy are presented in Section 8.2. The linear matrix inequalities (LMIs) based solution and controller reconstruction for this problem are given in Section 8.3. In Section 8.4, the implementation of the so-called "LPV semi-active" controller is described. Two controller parameterizations (one comfort oriented and the other road-holding oriented) and linear results are provided in Section 8.5. Nonlinear time and frequency simulations on the quarter vehicle model and performance index are evaluated in Section 8.6. Conclusions and potential perspectives are discussed in Section 8.7.

8.1 Synthesis Model

This section is devoted to the description of the models used for the design and synthesis of the "LPV semi-active" controller. First, the linear quarter model is recalled, and second, the considered static damper model is defined.

DOI: 10.1016/B978-0-08-096678-6.00008-0

8.1.1 System Model Σ_c

The control design is based on the linear vertical control oriented quarter-car model, denoted as (Σ_c), described by Definition 3.7, and recalled in equation (8.1).

$$\Sigma_c(c^0) := \begin{cases} M\ddot{z} = -k(z - z_t) - c^0(\dot{z} - \dot{z}_t) - F_d \\ m\ddot{z}_t = k(z - z_t) + c^0(\dot{z} - \dot{z}_t) + F_d - k_t(z_t - z_r) \\ \dot{F}_d = 2\pi\beta(F_{d_{in}} - F_d) \end{cases} \quad (8.1)$$

where k (resp. k_t) is the linearized stiffness coefficient of the suspension (resp. tire). Let z, z_t and F_d be the dynamics of the suspended mass, unsprung mass and actuator, respectively. Let denote $F_{d_{in}}$ as the control input. c^0 is a design parameter which represents the nominal suspension damping parameter; note that this parameter will play an important role in the design procedure.

The applied control law has the following structure:

$$F_{d_{in}} = u^{\mathscr{H}_\infty}(\rho) \quad (8.2)$$

where $u^{\mathscr{H}_\infty}(\rho)$ is the additional force provided by the controller, to be designed (see next sections). To account for actuator dissipative limitations and bounds (see Chapters 2 and 3), a new method is developed, based on the LPV polytopic theory using the $\mathscr{H}_\infty$ performance index.

8.1.2 Actuator Model

According to Definition 3.7, the considered semi-active damper will simply be modeled as a static map of the deflection speed/force of a controlled damper (see also Chapter 3), i.e. a lower and upper bound of the achievable forces, as given by Definition 8.1.

Definition 8.1 (*Controlled damper domain*). *Mathematically, the controlled damper domain or damper achievable force domain (denoted as $\mathscr{D}$ instead of $\mathscr{D}(c_{min}, c_{max}, c^0)$) may be described as follows (see also Figure 8.1):*

$$\mathscr{D}(c_{min}, c_{max}, c^0) := \left\{ U \in \mathbb{R} | \forall V \in \mathbb{R} : (U - (c_{max} - c^0)V)((c_{min} - c^0)V - U) \geq 0 \right\} \quad (8.3)$$

The $\mathscr{D}(c_{min}, c_{max}, c^0)$ domain is defined by the following parameters which depend on the considered controlled damper characteristics:

- c_{min}, is the minimal damping coefficient.
- c_{max}, is the maximal damping coefficient.
- c^0, is the nominal damping coefficient.

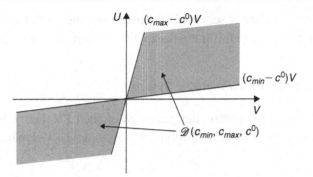

Figure 8.1: Dissipative domain $\mathscr{D}$ graphical illustration.

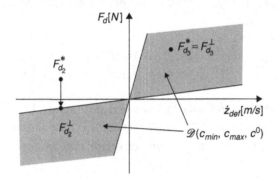

Figure 8.2: Clipping function illustration.

This static model is thus a saturation function of the deflection speed. Consequently, let us define the following clipping function $D(F_{d_{in}}, \dot{z}_{def})$ (Definition 8.2):

Definition 8.2 (*Clipping function*).   *Due to the controlled damper limitations (i.e. the effective force provided by the damper F_d should lie in the dissipative domain $\mathscr{D}$), the following clipping function $D(F_{d_{in}}, \dot{z}_{def})$ is defined (see also illustration in Figure 8.2):*

$$D : (F_{d_{in}}, \dot{z}_{def}) \mapsto F_{d_{in}} = \begin{cases} F_{d_{in}} & \text{if } F_{d_{in}} \in \mathscr{D} \\ F_{d_{in}}^{\perp} & \text{if } F_{d_{in}} \notin \mathscr{D} \end{cases} \qquad (8.4)$$

where $F_{d_{in}}$ is the required force (given by the controller) and $F_{d_{in}}^{\perp}$ is the orthogonal projection of $F_{d_{in}}$ on $\mathscr{D}$.

8.2 "LPV Semi-Active" Proposed Approach and Scheduling Strategy

In this section, the proposed "LPV semi-active" approach is described. Since this approach is based on a specific LPV modeling, some basic definitions concerning LPV models are recalled first. Then, the generalized control scheme is described together with the definition of the varying parameter (introduced to fulfill the dissipative property of the considered damper).

8.2.1 Basic Definition on LPV Polytopic Systems

Before presenting the proposed control design, based on the LPV theory, for the sake of completeness, the following general definitions of (polytopic) LPV dynamical systems are recalled.

Definition 8.3 (*LPV dynamical system*). *Given the linear matrix functions $A \in \mathbb{R}^{n \times n}$, $B \in \mathbb{R}^{n \times n_w}$, $C \in \mathbb{R}^{n_z \times n}$ and $D \in \mathbb{R}^{n_z \times n_w}$, a linear parameter varying (LPV) dynamical system (Σ_{LPV}) can be described as:*

$$\Sigma_{LPV} : \begin{cases} \dot{x}(t) = A(\rho(.))x(t) + B(\rho(.))w(t) \\ z(t) = C(\rho(.))x(t) + D(\rho(.))w(t) \end{cases} \tag{8.5}$$

where $x(t)$ is the state vector which takes values in a state space $X \in \mathbb{R}^n$, $w(t)$ is the input taking values in the input space $W \in \mathbb{R}^{n_w}$ and $z(t)$ is the output that belongs to the output space $Z \in \mathbb{R}^{n_z}$. Then, $\rho(.)$ is a varying parameter vector that takes values in the parameter space $\mathscr{P}_\rho$ such that,

$$\mathscr{P}_\rho := \left\{ \rho(.) := \begin{bmatrix} \rho_1(.) & \cdots & \rho_l(.) \end{bmatrix}^T \in \mathbb{R}^l \text{ and } \rho_i(.) \in \begin{bmatrix} \underline{\rho_i} & \overline{\rho_i} \end{bmatrix} \forall i = 1, \ldots, l \right\} \tag{8.6}$$

where l is the number of varying parameters. $\mathscr{P}_\rho$ is a convex set. For sake of readability, $\rho(.)$ will be denoted as ρ. Then, from a general viewpoint, when

- *$\rho(.) = \rho$, a constant value, (8.5) is a linear time invariant (LTI) system.*
- *$\rho(.) = \rho(t)$, where the parameter is a priori known, (8.5) is a linear time varying (LTV) system.*
- *$\rho(.) = \rho(x(t))$, (8.5) is a quasi-linear parameter varying (qLPV) system.*
- *$\rho(.) = \rho(\Omega)$, where Ω is an exogenous signal, (8.5) is an LPV system.*

Definition 8.4 (*Polytopic LPV dynamical system*). *An LPV system is said to be polytopic if it can be expressed as:*

$$\begin{bmatrix} A(\rho) & B(\rho) \\ \hline C(\rho) & D(\rho) \end{bmatrix} = \sum_{i=1}^{N} \alpha_i(\rho) \begin{bmatrix} A(\omega_i) & B(\omega_i) \\ \hline C(\omega_i) & D(\omega_i) \end{bmatrix} \in Co \left\{ \begin{bmatrix} A_1 & B_1 \\ \hline C_1 & D_1 \end{bmatrix}, \ldots, \begin{bmatrix} A_N & B_N \\ \hline C_N & D_N \end{bmatrix} \right\} \tag{8.7}$$

where $Co\{X\}$ denotes the convex hull of the $X \in V$ set of points (the convex hull or convex envelope for a set of points X in a real vector space V is the minimal convex set containing X).

ω_i are the vertices of the polytope formed by all the extremities of each varying parameter $\rho \in \mathscr{P}_\rho$, and $\alpha_i(\rho)$ defines a simplex as,

$$\alpha_i(\rho) := \frac{\prod_{k=1}^{l} |\rho_k - \mathscr{C}(\omega_i)_k|}{\prod_{k=1}^{l} (\overline{\rho}_k - \underline{\rho}_k)}, \; i = 1, \ldots, N \tag{8.8}$$

$$\alpha_i(\rho) \geq 0 \; and \; \sum_{i=1}^{N} \alpha_i(\rho) = 1 \tag{8.9}$$

where $\mathscr{C}(\omega_i)_k$ is the kth component of the vector $\mathscr{C}(\omega_i)$ defined as,

$$\mathscr{C}(\omega_i)_k := \{\rho_k | \rho_k = \overline{\rho}_k \; if \; (\omega_i)_k = \underline{\rho}_k \; or \; \rho_k = \underline{\rho}_k \; otherwise\} \tag{8.10}$$

Then, $N = 2^l$ is the number of vertices of the polytope formed by the extremum of each varying parameter ρ_i and A_i, B_i, C_i and D_i are constant known matrices (that represent the system evaluated at each vertex, i.e. $A_i = A(\omega_i)$, $B_i = B(\omega_i)$, $C_i = C(\omega_i)$ and $D_i = D(\omega_i)$.

For more details on LPV modeling, and on the properties of these systems, the reader is invited to refer to Biannic (1996) and Bruzelius (2004) PhD thesis (devoted to LPV modeling and control methods) and contributive work of Shamma and Athans (1991).

8.2.2 Generalized LPV Plant $\Sigma_g(\rho)$ and Problem Definition

As previously explained, we aim at designing a controller achieving $\mathcal{H}_\infty$ attenuation performance while guaranteeing the semi-active constraints. To do so, the generalized control scheme is chosen as in Figure 8.3, with the following blocks:

- $\Sigma_c(c^0)$, the linear quarter-car model used for the synthesis, as described in Definition 3.7 recalled in (8.1).
- The input weighting functions (W_i) are defined as:

$$W_i = \mathbf{diag}(W_{z_r}, W_n)$$
$$w = [z_r, n] \tag{8.11}$$

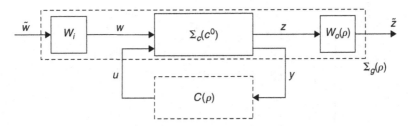

Figure 8.3: Generalized LPV scheme for the "LPV semi-active" control design.

where,

- W_{z_r} is used to shape the road disturbances effects z_r.
- W_n is used to handle the measurements noise n.

- The output weighting functions $(W_o(\rho))$ are defined as:

$$W_o(\rho) = \textbf{diag}(W_z, W_{z_t}, W_{F_d}(\rho))$$
$$z = [z, z_t, F_d]$$
(8.12)

where,

- W_z is used to weight the sprung mass behavior z (related to comfort performances).
- W_{z_t} is used to penalize the unsprung mass behavior z_t (related to road-holding performances).
- $W_{F_d}(\rho)$ is used to modify the control input signal F_d amplification. More specifically it is used to penalize (more or less) the control signal (according to the ρ signal). In the next sections, the effect of this weighting function to guarantee the semi-activeness is emphasized. In this particular case, it is convenient to define this filter as linearly dependent of ρ, the scheduling parameter which aims at guaranteeing the semi-activeness for the controller (see next subsection). The following choice is made:

$$W_{F_d}(\rho) = \rho W_{F_d}$$
(8.13)

where $\rho \in [\underline{\rho}; \overline{\rho}] \subseteq \mathbb{R}^+$ and W_{F_d} is a strictly proper LTI filter. Since the design to be performed is made in the $\mathscr{H}_\infty$ framework, the following remarks hold:

- when ρ is high, $W_{F_d}(\rho)$ is "large", therefore, it tends to attenuate the F_d signal.
- when ρ is low, $W_{F_d}(\rho)$ is "small", therefore, it does not attenuate the F_d signal.

Definition 8.5 ("LPV semi-active" generalized system form). *Let us consider the system interconnection described in Figure 8.3, where the system $\Sigma_c(c^0)$ is described by (8.1), the weighting filters W_i and $W_o(\rho)$ are described as (8.11) and (8.12), then the "LPV semi-active" generalized system (Σ_g) is given as,*

$$\Sigma_g : \begin{cases} \dot{\xi}(t) = A(\rho)\xi(t) + B_1(\rho)\tilde{w}(t) + B_2 u(t) \\ \tilde{z}(t) = C_1(\rho)\xi(t) + D_{11}(\rho)\tilde{w}(t) + D_{12}u(t) \\ y(t) = C_2\xi(t) \end{cases}$$
(8.14)

where,

$$\begin{cases} \xi(t) = \begin{bmatrix} x_{\Sigma_c}(t) & x_w(t) \end{bmatrix}^T \\ \tilde{z}(t) = \begin{bmatrix} W_z z(t) & W_{z_t} z_t(t) & W_{F_d}(\rho) F_d(t) \end{bmatrix}^T \\ \tilde{w}(t) = \begin{bmatrix} W_{z_r}^{-1} z_r(t) & W_n^{-1} n(t) \end{bmatrix}^T \\ y(t) = z_{def}(t) \\ u(t) = F_{d_{in}}(t) \\ \rho \in \begin{bmatrix} \underline{\rho} & \overline{\rho} \end{bmatrix} \end{cases}$$
(8.15)

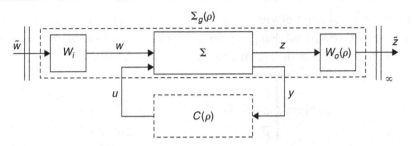

Figure 8.4: Generalized $\mathcal{H}_\infty$ control scheme.

where $\xi(t)$ is the concatenation of the state $x_{\Sigma_c}(t)$ (of the linearized quarter vehicle model (8.1)) and of the weighting function state variables, which takes its values in $\Xi \in \mathbb{R}^n$, $\tilde{z}(t)$ the performance output which takes its values in $Z \in \mathbb{R}^{n_z}$, $\tilde{w}(t)$ the weighted input which takes its values in $W \in \mathbb{R}^{n_w}$, $y(t)$ the measured signal which takes its values in $Y \in \mathbb{R}^{n_y}$, $u(t)$ the control signal which takes its values in $U \in \mathbb{R}^{n_u}$ and $\rho \in \mathcal{P}_\rho$ the varying parameter. Then, $A(\rho) \in \mathbb{R}^{n \times n}$, $B_1(\rho) \in \mathbb{R}^{n \times n_w}$, $B_2 \in \mathbb{R}^{n \times n_u}$, $C_1(\rho) \in \mathbb{R}^{n_z \times n}$, $D_{11}(\rho) \in \mathbb{R}^{n_z \times n_w}$, $D_{12} \in \mathbb{R}^{n_z \times n_u}$ and $C_2 \in \mathbb{R}^{n_y \times n}$ are known matrices. ρ is assumed to be known and measurable.

Remark 8.1 (*Parametrization of W_i and $W_o(\rho)$*). *Since these weighting filters determine the obtained closed-loop performances, i.e. to achieve either comfort or road-holding performances, they will be more precisely described in Section 8.5 as study cases. Note also that, although the system $(\Sigma_c(c^0))$ is LTI, the generalized system (Σ_g) is LPV since at least one weighting filter is parameter dependent (here W_{F_d}).*

In the $\mathcal{H}_\infty$ approach, the control synthesis relies on a disturbance attenuation problem. It consists in finding a stabilizing controller that minimizes the impact of the input disturbances $\tilde{w}(t)$ on the controlled output $\tilde{z}(t)$. In the case of the LPV $\mathcal{H}_\infty$ control, this impact must be measured thanks to the induced $\mathcal{L}_2$-$\mathcal{L}_2$ norm, which is referred to as a $\mathcal{H}_\infty$ problem, and is represented in Figure 8.4. The solution to this problem is described in the next section.

8.2.3 Scheduling Strategy for the Parameter ρ

The underlying idea of the "LPV semi-active" control design is to increase the control gain $F_{d_{in}}$ when the required force belongs to the allowed semi-active domain $\mathcal{D}$, and otherwise to rely on the nominal damping when the forces are outside the allowable space. To satisfy the dissipative damper constraints (see next subsection), the ρ parameter is tuned in a particular way:

- when ρ is high, $W_{F_d}(\rho)$ is "large", therefore, it tends to attenuate the F_d signal.
- when ρ is low, $W_{F_d}(\rho)$ is "small", therefore, it does not attenuate the F_d signal.

Therefore, from a general point of view, ρ may be viewed as an anti-windup signal, computed on the actuator model (controlled damper model), and is similar to a variable saturation. For that purpose, the following scheduling strategy $\rho(\varepsilon)$ is introduced:

$$\rho(\varepsilon) := \begin{cases} \underline{\rho} & \text{if } \varepsilon < \mu \\ \underline{\rho} + \frac{\overline{\rho} - \underline{\rho}}{\mu}(\varepsilon - \mu) & \text{if } \mu \le \varepsilon \le 2\mu \\ \overline{\rho} & \text{if } \varepsilon > 2\mu \end{cases} \qquad (8.16)$$

$$\varepsilon = ||F_{d_{in}} - F_{d_{in}}^{\perp}||_2 \qquad (8.17)$$

where ε is the distance between the required force and the force projected on domain $\mathscr{D}$ (according to the function $D(F_{d_{in}}, \dot{z}_{def})$). μ is a design parameter that modifies the dead-zone of the $\rho(\varepsilon)$ function. The parameter μ is theoretically zero but, for numerical reasons, it is chosen sufficiently low (e.g. $\mu = 0.1$) to ensure the semi-active control. Note that this choice is reasonable to guarantee semi-active constraint, and makes the problem numerically tractable (see Section 8.6).

Remark 8.2 (*About the scheduling parameters $\rho(\varepsilon)$*)

- *In (8.16) $\rho(\varepsilon)$ belongs to a closed set $[\underline{\rho}, \overline{\rho}]$ which is essential in the LPV framework.*
- *$\varepsilon \ne 0 (\Leftrightarrow F_{d_{in}} \ne F_{d_{in}}^{\perp})$ means that the required force is outside the allowed range. Conversely, $\varepsilon = 0 (\Leftrightarrow F_{d_{in}} = F_{d_{in}}^{\perp})$ means that the force required by the controller is reachable for the considered semi-active actuator.*
- *Note that the scheduling parameter is very similar to what is done in the anti-windup compensator synthesis literature, i.e. it represents the saturation as a dead-zone function (see e.g. Roos, 2007). Here, since the saturation depends on the system states, the dead-zone is time varying.*

8.3 $\mathcal{H}_{\infty}$ LMI Based "LPV Semi-Active" Controller Synthesis

The idea is to synthesize a parameter dependent controller $C(\rho)$ in order to satisfy some performance objectives while ensuring the semi-activeness. In other words, the objective is to add a force ($u^{\mathcal{H}_{\infty}}$) when required and possible, and to rely on the nominal solution (defined by c^0) when no force can be added. For that purpose, the LPV synthesis framework is used and the performances are described through $\mathcal{H}_{\infty}$ performance criteria (to ensure some "robustness" properties). In this section, the $\mathcal{H}_{\infty}$ LMI solution of the "LPV semi-active" control problem is described in two steps:

1. Problem feasibility and Controller reconstruction.
2. Numerical issues and "LPV semi-active" synthesis algorithm.

8.3.1 Problem Feasibility and Controller Reconstruction

In the generalized LPV plant presented in Figure 8.3, $W_{F_d}(\rho)$, the performance criterion for the control signal, is ρ dependent. Let us recall that in the $\mathcal{H}_\infty$ framework, this weight indicates how large the gain on the control signal can be. The usual rule of thumb is:

- Choosing a large ρ forces the control signal to be small. When ρ is the largest, the control signal is so penalized that it is practically zero, and the control signal would simply be $F_{d_{in}} = 0$.
- Choosing a small ρ allows the control signal to be large. In this case, the control signal is no longer penalized and the control signal would be $F_{d_{in}} = u^{\mathcal{H}_\infty}(\rho)$.

The $\mathcal{H}_\infty$ control synthesis solution for LPV systems is extended from the LTI system. Both following propositions solve the $\mathcal{H}_\infty$ "LPV semi-active" problem using a polytopic approach (the first one is related to the feasibility, and the second one, to controller reconstruction) (Scherer et al., 1997).

Proposition 8.1 (*Feasibility –* $\mathcal{H}_\infty$ *LMI based "LPV semi-active"*).   *Let us consider the interconnection in Figure 8.4, where $\Sigma_g(\rho)$ is defined by the state space representation given in (8.14). There exists a full order dynamical output feedback controller of the form,*

$$C(\rho): \begin{cases} \dot{x}_c(t) = A_c(\rho)x_c(t) + B_c(\rho)y(t) \\ u(t) = C_c(\rho)x_c(t) \end{cases} \qquad (8.18)$$

that minimizes the LPV polytopic $\mathcal{H}_\infty$ norm if there exist symmetric matrices $X, Y \in \mathbb{R}^{n\times n}$, and matrices $\widetilde{A}(\underline{\rho}), \widetilde{A}(\overline{\rho}) \in \mathbb{R}^{n\times n}$, $\widetilde{B}(\underline{\rho}), \widetilde{B}(\overline{\rho}) \in \mathbb{R}^{n\times n_u}$, $\widetilde{C}(\underline{\rho}), \widetilde{C}(\overline{\rho}) \in \mathbb{R}^{n_y\times n}$ and $\gamma \in \mathbb{R}^{+*}$ that are solutions to the following problem:

$$\gamma^* = \min \ \gamma$$
$$s.t. \ (8.20) \qquad (8.19)$$
$$s.t. \ (8.21)$$

$$\begin{bmatrix} AX + B_2\widetilde{C}(\underline{\rho}) + (\star)^T & (\star)^T & (\star)^T & (\star)^T \\ \widetilde{A}(\underline{\rho}) + A^T & YA + \widetilde{B}(\underline{\rho})C_2 + (\star)^T & (\star)^T & (\star)^T \\ B_1^T & B_1^T Y + D_{21}^T\widetilde{B}(\underline{\rho})^T & -\gamma I & (\star)^T \\ C_1 X + D_{12}\widetilde{C}(\underline{\rho}) & C_1 & D_{11} & -\gamma I \end{bmatrix} \prec 0$$

$$\begin{bmatrix} AX + B_2\widetilde{C}(\overline{\rho}) + (\star)^T & (\star)^T & (\star)^T & (\star)^T \\ \widetilde{A}(\overline{\rho}) + A^T & YA + \widetilde{B}(\overline{\rho})C_2 + (\star)^T & (\star)^T & (\star)^T \\ B_1^T & B_1^T Y + D_{21}^T\widetilde{B}(\overline{\rho})^T & -\gamma I & (\star)^T \\ C_1 X + D_{12}\widetilde{C}(\overline{\rho}) & C_1 & D_{11} & -\gamma I \end{bmatrix} \prec 0$$

$$(8.20)$$

$$\begin{bmatrix} X & I \\ I & Y \end{bmatrix} \succ 0 \qquad (8.21)$$

where $\widetilde{A}(\rho)$, $\widetilde{B}(\rho)$, $\widetilde{C}(\rho)$, $\widetilde{A}(\overline{\rho})$, $\widetilde{B}(\overline{\rho})$, $\widetilde{C}(\overline{\rho})$, X and Y are the decision variables.

Proposition 8.2 (*Reconstruction – $\mathcal{H}_\infty$ LMI based "LPV semi-active"*). *If such a controller $C(\rho)$ exists (Feasibility Proposition 8.1), the controller reconstruction is obtained by solving the following system of equations at each vertex of the polytope, i.e.:*

$$\begin{aligned} solve \quad & (8.23)\,|_{\rho=\underline{\rho}} \\ & (8.23)\,|_{\rho=\overline{\rho}} \end{aligned} \qquad (8.22)$$

$$\begin{cases} C_c(\rho) = \widetilde{C}(\rho)M^{-T} \\ B_c(\rho) = N^{-1}\widetilde{B}(\rho) \\ A_c(\rho) = N^{-1}(\widetilde{A}(\rho) - YAX - NB_c(\rho)C_2X - YB_2C_c(\rho)M^{-T})M^{-T} \end{cases} \qquad (8.23)$$

where M and N are defined such that $MN^T = I - XY$ which are chosen by applying a singular value decomposition and a Cholesky factorization, as follows:

1. *Singular value decomposition: $I - XY = U\Sigma V^T$ (where U and V are unitary matrices, and Σ, a diagonal matrix).*
2. *Cholesky factorization: $\Sigma = R^T R$ (where R is a real upper triangular matrix). Therefore, $I - XY = URR^T V^T$.*
3. *Then one can choose, $M = UR^T$ and $N = VR^T$.*

For more details on the LMI, the reader is invited to refer to the contributive work of Apkarian and Gahinet (1995); Chilali et al. (1999); Scherer et al. (1997). Note that proofs and comments are also provided in Poussot-Vassal (2008) PhD thesis.

8.3.2 Numerical Issues and "LPV Semi-Active" Synthesis Algorithm

The previous propositions are sufficient to ensure the feasibility and the reconstruction of the "LPV semi-active" controller. Here, some additional practical issues are given.

Remark 8.3 (*Conditioning improvement*). *For practical numerical issues, LMIs (8.20) should be solved firstly to find γ^*, the optimal bound solution. Then, LMIs (8.20) may be solved again by fixing the attenuation level $\gamma = \gamma^*(1 + v/100)$, where $v > 0$ may be viewed as a percentage. In this second step, inequality (8.21) is replaced by,*

$$\begin{bmatrix} X & \alpha I \\ \alpha I & Y \end{bmatrix} \succ 0 \qquad (8.24)$$

where $\alpha > 0$. Then, the optimization to be done consists in maximizing α. This procedure maximizes the minimal eigenvalue of XY, which is then pushed away from I. This avoids bad conditioning when inverting M and N in the controller reconstruction step (Scherer et al., 1997).

According to these Propositions and Remarks, Algorithm 3 may be used for the synthesis of the $\mathcal{H}_\infty$ "LPV semi-active" suspension controller (it summarizes all the procedure).

Algorithm 3 $\mathcal{H}_\infty$ "LPV Semi-Active" Suspension Controller Design

---

1. Polytopic system definition:
   According to Section 8.2, construct the LPV polytopic system, i.e. $\Sigma_g(\rho)$ and $\Sigma_g(\overline{\rho})$ such that

   $$\Sigma_g(\rho) \in \mathbf{Co}\{\Sigma_g(\underline{\rho}), \Sigma_g(\overline{\rho})\} \tag{8.25}$$

   Note that to complete this step, weighting filters should be defined and parameterized (refer to Section 8.5 for details).

2. Feasibility – $\mathcal{H}_\infty$ LMI based "LPV semi-active":

   $$\begin{aligned} \gamma^* = \min \quad & \gamma \\ & \text{s.t. } (8.20) \\ & \text{s.t. } (8.21) \end{aligned} \tag{8.26}$$

   and find γ^*, the optimal $\mathcal{H}_\infty$ bound.

3. Conditioning improvement: fix $\gamma > \gamma^*$, compute

   $$\begin{aligned} \max \quad & \alpha \\ & \text{s.t. } (8.20) \\ & \text{s.t. } (8.24) \end{aligned} \tag{8.27}$$

   and find $\widetilde{\mathbf{A}}(\underline{\rho}), \widetilde{\mathbf{B}}(\underline{\rho}), \widetilde{\mathbf{C}}(\underline{\rho}), \widetilde{\mathbf{A}}(\overline{\rho}), \widetilde{\mathbf{B}}(\overline{\rho}), \widetilde{\mathbf{C}}(\overline{\rho}), \mathbf{X}$ and $\mathbf{Y}$.

4. Reconstruction – $\mathcal{H}_\infty$ LMI based "LPV semi-active":
   - Find the appropriate matrices M and N, e.g. by singular values decomposition and Cholesky factorization of X and Y.
   - Reconstruct the controller $C(\rho)$ at each vertex by solving:

   $$\begin{aligned} \text{solve} \quad & (8.23) \,|_{\rho=\underline{\rho}} \\ & (8.23) \,|_{\rho=\overline{\rho}} \end{aligned} \tag{8.28}$$

5. Result: The two dynamical full order controllers (8.29) are the solution of the Robust $\mathcal{H}_\infty$ "LPV semi-active" control problem.

   $$\begin{aligned} C(\underline{\rho}) &= \begin{bmatrix} A_c(\underline{\rho}) & B_c(\underline{\rho}) \\ C_c(\underline{\rho}) & 0 \end{bmatrix} \\ C(\overline{\rho}) &= \begin{bmatrix} A_c(\overline{\rho}) & B_c(\overline{\rho}) \\ C_c(\overline{\rho}) & 0 \end{bmatrix} \end{aligned} \tag{8.29}$$

6. Apply control the law using equation (8.30), see also Figure 8.5

In the next section, the implementation scheme of this controller is recalled.

8.4 Controller Implementation and On-Line Scheduling

Applying the LMI based polytopic LPV $\mathcal{H}_\infty$ control synthesis (described in Section 8.3) to the generalized plant (8.14) leads to two controllers $C(\underline{\rho})$ and $C(\overline{\rho})$, hence to two closed-loops ($CL(\underline{\rho})$ and $CL(\overline{\rho})$). The applied control law is the convex combination of these two controllers, function of ρ, as described in equation (8.30),

$$F_{d_{in}} = u^{\mathcal{H}_\infty}(\rho)$$

$$u^{\mathcal{H}_\infty}(\rho) = \left[\frac{|\rho-\overline{\rho}|}{\overline{\rho}-\underline{\rho}}C(\underline{\rho}) + \frac{|\rho-\underline{\rho}|}{\overline{\rho}-\underline{\rho}}C(\overline{\rho})\right]z_{def}$$

(8.30)

Hence, the controller $C(\rho)$ and the closed-loop $CL(\rho)$ lie in the following convex hulls:

$$C(\rho) \in \mathbf{Co}\{C(\underline{\rho}), C(\overline{\rho})\}$$

$$CL(\rho) \in \mathbf{Co}\{CL(\underline{\rho}), CL(\overline{\rho})\}$$

(8.31)

Before providing the simulation results, a sketch of the implementation scheme for the proposed strategy is given. Figure 8.5 shows the implementation synopsis, for any kind of semi-active actuators.

where,

- "SA mdl" is the model of the semi-active damper considered in the application (here $D(u, \dot{z}_{def})$, a static model of the upper and lower damping factors of the controlled damper, as described in 8.1).
- $\rho(\varepsilon)$ is the scheduling law given in (8.16).
- "SA act" is the real considered controlled damper actuator that modifies the damping coefficient.
- $F_{d_{in}}$, F_d, ε and ρ are the control and scheduling signals as described before.
- Ω is used here to specify the set of input parameters of the real controllable damper "SA act". As an illustration, for a MR damper, the real control input is I_{MRD}, the current that modifies the magnetorheological fluid viscosity. Hence, in the case of the MR damper actuator, the force $F_{d_{in}}$ has to be converted into a current (by the means of tables or electrical laws for instance), hence $\Omega = I_{MRD}$. For other kinds of semi-active actuators, control input can be some mechanical, pneumatic elements or other damper input variables depending on the chosen technology (see e.g. Delphi, 2008; Lord, 2008; Sachs, 2008), and Chapter 2.

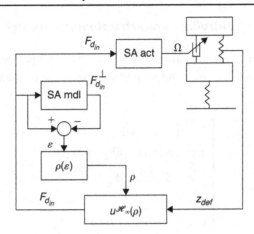

Figure 8.5: Implementation scheme.

Remark 8.4 (*About the implementation*)

- *This scheme is the same for any suspension control approach. The advantage here is that the required force $F_{d_{in}}$ always remains in the allowed force range of the considered controlled semi-active actuator (thanks to the proposed LPV approach).*

- *If a dynamical model of the considered actuator (gain, bandwidth limit...) is available, it may be included in the synthesis by the mean of the $W_{F_d}(\rho)$ weighting function.*

8.5 Controller Parametrization

In the control design step, the obtained performances highly depend on the weighting function definitions (W_i and $W_o(\rho)$). Only the description of the control problem and its solution have been developed so far. In the next subsections, two illustrative examples of controller parametrization, i.e. weighting function definitions, are presented:

1. A comfort oriented one (focusing on the z behavior).
2. A road-holding oriented one (focusing on the z_t behavior).

Both are easy to adapt to any quarter-car model. After having presented the synthesis results of these controllers (obtained thanks to the LMI resolution presented before), nonlinear time and frequency simulations are shown to illustrate the efficiency of the design. In this section, the motorcycle parameters given in Spelta (2008) and in Table 1.2 are used.

Remark 8.5. *There is nearly an infinite number of ways to parametrize the performance weight. In the following, two weighting configurations are presented to enhance either comfort or road-holding performances. These choices may be improved by adjusting the weighting function descriptions.*

8.5.1 Comfort Oriented Controller Parametrization (Controller 1)

To achieve comfort specification, one aims at penalizing the suspended mass behavior (z). Then, the following controlled output and weight functions are chosen:

$$
\begin{cases}
\xi = \begin{bmatrix} x_{\Sigma_c} & x_w \end{bmatrix}^T \\
\tilde{z} = \begin{bmatrix} W_z(\rho)z & W_{F_d}(\rho)F_d \end{bmatrix}^T \\
\tilde{w} = \begin{bmatrix} W_{z_r}^{-1}z_r & W_n^{-1}n \end{bmatrix}^T \\
y = z_{def} \\
\rho \in \begin{bmatrix} \underline{\rho} & \overline{\rho} \end{bmatrix} = \begin{bmatrix} 0.01 & 10 \end{bmatrix}
\end{cases}
\tag{8.32}
$$

where,

- $c^0 = 1500$ is used as the nominal damping factor. This choice is motivated by a comfort oriented objective.
- $W_{z_r} = 10^{-3}$ is used to shape the road disturbances magnitudes. In this case, magnitudes are shaped to 1 cm. This consideration is essential for synthesis purposes, in order to reduce conservatism in the LMIs resolution.
- $W_n = 10^{-1}$ is used to take into consideration the noise that can enter on the measure signal (here, 10% is considered).
- $W_z(\rho) = (1 - \rho/\overline{\rho})G_z\frac{s/(2\pi f_{11})+1}{s/(2\pi f_{12})+1}$ is used to penalize the specific suspended mass (M) frequencies when ρ is low. In our case, the objective is to enhance comfort, therefore, one aims at attenuating the first resonance peak of the function $\frac{z}{z_r}$. To do so, one chooses $f_{11} = 2\,\text{Hz}$ and $f_{12} = \frac{1}{2\pi}\sqrt{\frac{k_t}{m_t}}\,\text{Hz}$ (which is an invariant point, see Properties 3.1, 3.2 and 3.3).
- $W_{F_d}(\rho) = \rho\frac{1}{s/(5\beta)+1}$ is used to achieve the semi-activeness property (where β is defined in (8.1)). The low pass shape of this filter is designed to limit attenuation over higher frequencies (transfer between F_d and $W_{F_d}(\rho)F_d$).

After synthesis (performed according to the Algorithm 3 described below), by using YALMIP (Lofberg, 2004) LMI parser together with SeDuMi (Sturm, 1999) semi-definite problem solver, the following Bode diagrams and γ attenuation bound are obtained (see Figure 8.6).

From Figure 8.6, it appears that:

- When ρ is small, the frequencies of interest of the transfer F_z are globally improved (Figure 8.6 – top), while the transfer z_t/z_r is deteriorated (Figure 8.6 – bottom).
- When ρ is large, the closed-loop behavior is similar to the passive suspension with the nominal damping factor (here $c^0 = 1500$).

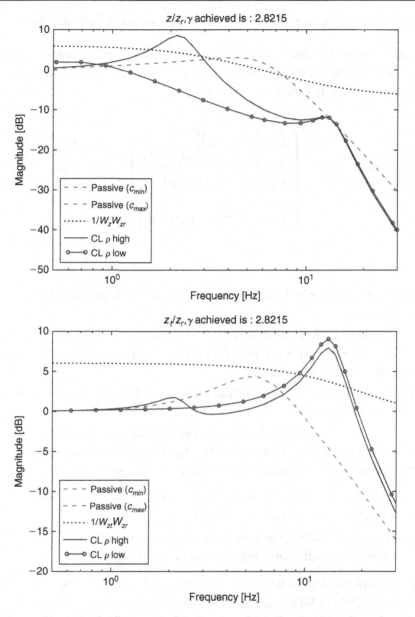

Figure 8.6: Controller 1: Bode diagrams of F_z (top) and F_{z_t} (bottom), evaluated at each vertex of the polytope. (Here, the c_{max} Bode response coincides with ρ "high").

Consequently, when the provided controller force lies in the achievable controlled damper domain $\mathcal{D}$, the quarter-car comfort behavior tends to be improved (while deteriorating the road-holding performances). But when the controller force is outside of $\mathcal{D}$, the controller would provide a nominal damping factor, in order to turn back inside the allowed domain.

8.5.2 Road-Holding Oriented Controller Parametrization (Controller 2)

Similarly to the previous comfort oriented parametrization, to achieve road-holding specification, one aims at penalize the unsprung mass behavior (z_t). Then, the following controlled output and weight functions are chosen:

$$
\begin{cases}
\xi = \begin{bmatrix} x_{\Sigma_c} & x_w \end{bmatrix}^T \\
\tilde{z} = \begin{bmatrix} W_{z_t} z_t(\rho) & W_{F_d}(\rho) F_d \end{bmatrix}^T \\
\tilde{w} = \begin{bmatrix} W_{z_r}^{-1} z_r & W_n^{-1} n \end{bmatrix}^T \\
y = z_{def} \\
\rho \in \begin{bmatrix} \underline{\rho} & \overline{\rho} \end{bmatrix} = \begin{bmatrix} 0.01 & 10 \end{bmatrix}
\end{cases}
\tag{8.33}
$$

where,

- $c^0 = 3000$ is used as nominal damping factor. This choice is motivated by a road-holding oriented objective.

- $W_{z_r} = 10^{-2}$ is used to shape road disturbance magnitudes. In this case, they are shaped to 1 cm. This consideration is essential for synthesis purposes, in order to reduce conservatism in the LMIs resolution.

- $W_n = 10^{-1}$ is used to take into consideration the noise that can enter on the measure signal (here, 10% is considered).

- $W_{z_t}(\rho) = (1 - \rho/\overline{\rho}) G_{z_t} \frac{s/(2\pi f_{21})+1}{s/(2\pi f_{22})+1}$ is used to penalize the specific unsprung mass (m) frequencies (when ρ is low). In our case, the objective is to enhance road-holding, therefore, one aims at attenuating the first resonance peak of the function $\frac{z_t}{z_r}$. To do so, one chooses $f_{21} = \frac{1}{2\pi} \sqrt{\frac{k_t}{m_t}}$ Hz and $f_{22} = \frac{1}{2\pi} \sqrt{\frac{kt}{ms+mus}}$ Hz (which are invariant points of the quarter-car model, see Properties 3.1, 3.2 and 3.3).

- $W_{F_d}(\rho) = \rho \frac{1}{s/(5\beta)+1}$ is used to achieve the semi-active property. The low pass design of this filter ensures attenuation over higher frequencies (transfer between F_d and $W_{F_d}(\rho) F_d$).

After synthesis (performed according to the Algorithm 3 described below), by using YALMIP (Lofberg, 2004) LMI parser together with SeDuMi (Sturm, 1999) semi-definite problem solver, the following Bode diagrams and γ attenuation bound are obtained (see Figure 8.7),

From Figure 8.7, it appears that:

- When ρ is low, the frequencies of interest of the transfer F_{z_t} are globally improved (Figure 8.7 – bottom), while the transfer F_z is deteriorated (Figure 8.7 – top).
- When ρ is high, the closed-loop behavior is similar to the passive suspension with the nominal damping factor (here $c^0 = 3000$).

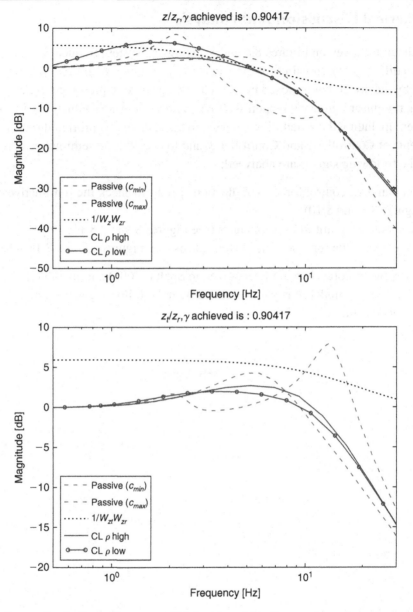

Figure 8.7: Controller 2: Bode diagrams of F_z (top) and F_{z_t} (bottom), evaluated at each vertex of the polytope.

Similarly to the previous comfort oriented case (Controller 1), when the provided controller force lies in the achievable controlled damper domain $\mathcal{D}$, the quarter-car road-holding behavior tends to be improved (while deteriorating the comfort performances). But when the controller force is outside of $\mathcal{D}$, the controller would provide a nominal damping factor, in order to turn back inside the allowed domain.

8.6 Numerical Discussion and Analysis

The Bode diagrams given on Figures 8.6 and 8.7 show the system frequency behavior according to different frozen values of the parameter ρ (for $\rho = \underline{\rho}$ and $\rho = \overline{\rho}$). Since the ρ is not chosen by the driver, but imposed by the scheduling strategy presented in Section 8.2, in this section, nonlinear frequency responses (FR) and performance evaluations are performed on the system including the scheduling strategy, to evaluate the comfort and road-holding improvements of Controller 1 and Controller 2, and to check if the semi-active constraint is fulfilled. The following signals are analyzed:

- SER diagram (i.e. control force vs. deflection speed), to check the semi-active constraint (see Figures 8.8 and 8.10).
- $\tilde{F}_z$, to evaluate the comfort performances (see Figures 8.9 and 8.11 – top).
- $\tilde{F}_{z_{def_t}}$, to evaluate the road-holding performances (see Figures 8.9 and 8.11 – bottom).

In this section, the motorcycle parameters given in Spelta (2008) and in Table 1.2 are used. Moreover, the damper model $\mathscr{D}$ is parameterized with the following values: $c_{min} = 900\,\text{Ns/m}$ and $c_{max} = 4300\,\text{Ns/m}$.

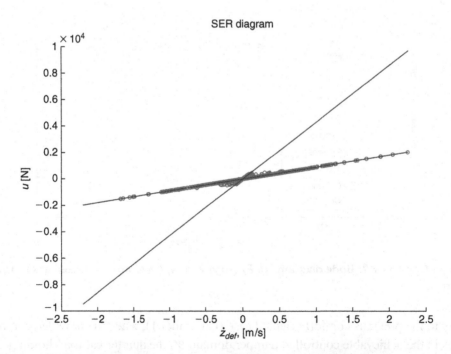

Figure 8.8: Controller 1: Force vs. Deflection speed diagram of the frequency response (with $z_r = 5\,\text{cm}$ from 1 to 20 Hz). "LPV semi-active" comfort oriented (round symbols), $c_{min} = 900\,\text{Ns/m}$ and $c_{max} = 4300\,\text{Ns/m}$ limits (solid lines).

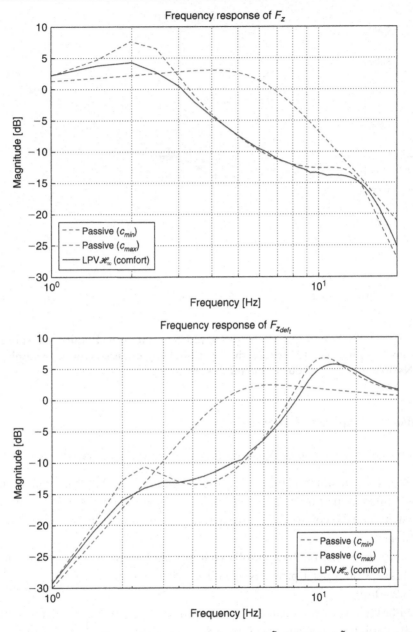

Figure 8.9: Controller 1: Frequency responses of $\tilde{F}_z$ (top) and $\tilde{F}_{z_{def_t}}$ (bottom).

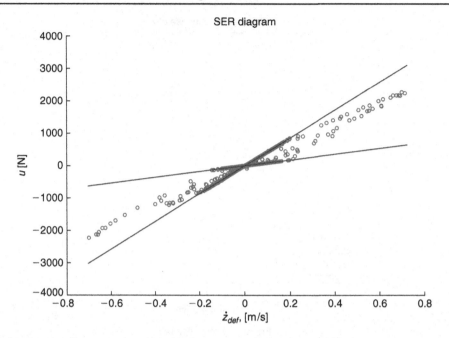

Figure 8.10: Controller 2: Force vs. deflection speed diagram of the frequency response (with $z_r = 5\,\text{cm}$ **from 1 to 20 Hz). "LPV semi-active" road-holding oriented (round symbols),** $c_{min} = 900\,\text{Ns/m}$ **and** $c_{max} = 4300\,\text{Ns/m}$ **limits (solid lines).**

8.6.1 Nonlinear Frequency Response

The following observations can be made for this first controller parametrization:

1. The semi-active constraint is fulfilled as illustrated in Figure 8.8. Note that in this figure, the most important part of the provided forces is related to low damping values, which is consistent with the comfort objective and the attenuation of the suspended mass oscillations.
2. The comfort properties on the controlled damper are improved compared to the passive nominal case. Controller 1 provides a good attenuation of the F_z transfer function (see Figure 8.9 – top).
3. The road-holding performances are deteriorated: the $\tilde{F}_{z_{def}}$ transfer is deteriorated especially in high frequencies (see Figure 8.9 – bottom).

Similarly to the first parametrization, the following remarks and comments can be made on the second controller parametrization:

1. The semi-active constraint is fulfilled as illustrated in Figure 8.10. Note that in this figure, the most important part of the provided forces is equivalent to high damping values, which is consistent with the idea of road-holding and minimization of tire deflection objective.

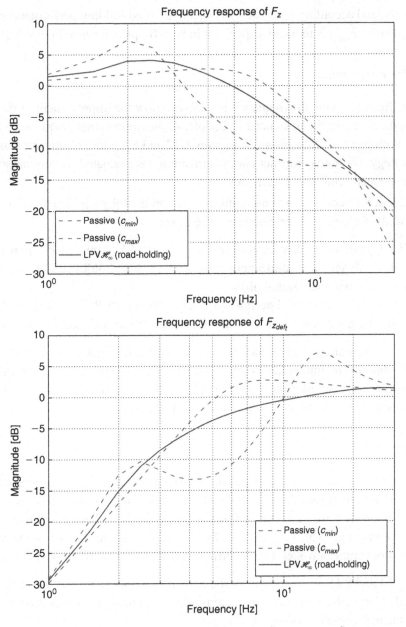

Figure 8.11: Controller 2: Frequency responses of $\tilde{F}_z$ (top) and $\tilde{F}_{z_{def_t}}$ (bottom).

2. The comfort properties on the controlled damper are deteriorated compared to the passive nominal case. Controller 2 provides a poor attenuation of the $\tilde{F}_z$ transfer function (see Figure 8.11 – top).

3. Conversely, and according to the design objectives, road-holding performances are greatly enhanced since $\tilde{F}_{z_{def_t}}$ is reduced, especially in high frequencies (see Figure 8.11 – bottom).

8.6.2 Performance Index

The previous figures, giving the frequency responses, show the improvement of the approach. Using the evaluation criteria introduced in Chapter 4, the improvement is measured by comparison with previous algorithms. In Figures 8.12 and 8.13, the improvement of the proposed strategy (for both controller parameterizations) is evaluated by comparison with previously presented semi-active strategies.

Figures 8.12 and 8.13 confirm the observations made on the FR plots. Then, from a global point of view, the following observations may be made:

- Controller 1 (resp. Controller 2), designed using the "LPV semi-active" approach shows how to efficiently enhance comfort criterion J_c (resp. road-holding criterion J_{rh}) according to the parametrization objectives.
- Compared to the other control design (see Chapter 6), the "LPV semi-active" Controllers 1 and 2 provide good performance results for both comfort and road-holding objectives.

The reader may also notice that the main interest lies in the fact that intermediate performance can be easily obtained by tuning the weighting filters or by designing a new controller with an additional performance oriented parameter in order to define a convex combination of the 2 controller parametrization sets, achieving mixed performances.

8.6.3 Bump Test

The bump test refers to the classical response to a triangular bump in the road profile (Figure 4.7). This kind of excitation is very realistic since it means going through a usual 6 cm road bump at a speed of 30 km/h. On the following figures, the "LPV semi-active" strategy is evaluated and compared to the Passive case. More precisely, Controller 1 parametrization (comfort oriented) and Controller 2 parametrization (road-holding oriented) are compared.

On Figures 8.14 and 8.15, the responses of the body displacement and of the suspension stroke are displayed. Also in this case, the conclusions which can be drawn are consistent with the previous frequency-domain analysis:

- "LPV semi-active" Controller 1 parametrization provides very good damping, and filtering of the bump, while "LPV semi-active" Controller 2 shows the worst signal amplification.
- Conversely, Controller 2 shows how to limit the suspension deflection amplification, enhancing the road-holding behavior.

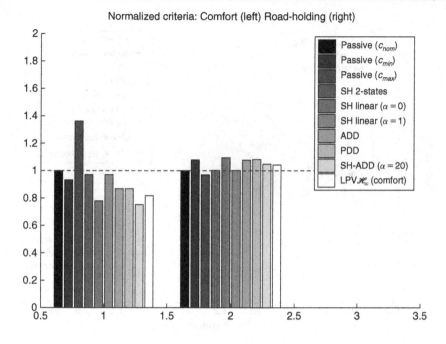

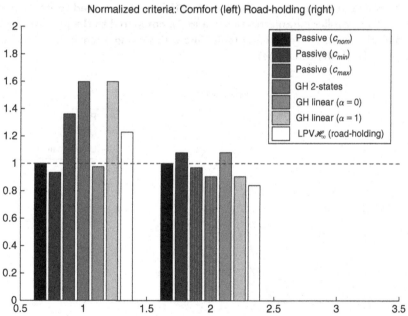

Figure 8.12: Normalized performance criteria comparison: comfort oriented "LPV semi-active" design compared to other comfort oriented control laws (top) and road-holding oriented "LPV semi-active" design compared to other road-holding control laws (bottom). Comfort criterion – J_c **(left histogram set) and road-holding criterion –** J_{rh} **(right histogram set).**

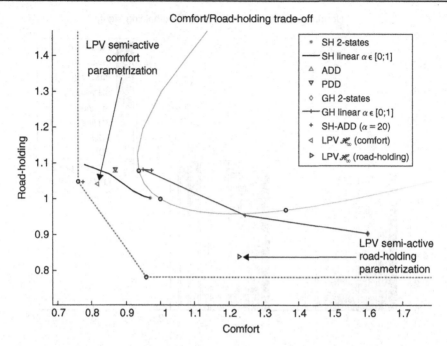

Figure 8.13: Normalized performance criteria trade-off for the presented control algorithms and "LPV semi-active" (controller parametrization 1 and 2), compared to the passive suspension system, with damping value $c \in [c_{min}; c_{max}]$ (solid line with varying intensity), optimal comfort and road-holding bounds (dash dotted line).

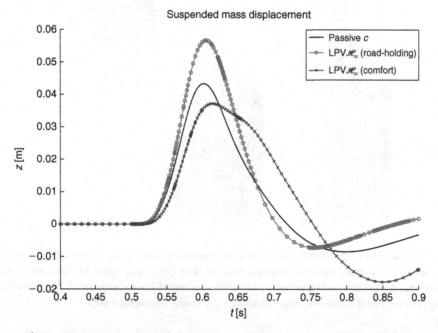

Figure 8.14: Bump test: Time response of chassis z – comfort criterion.

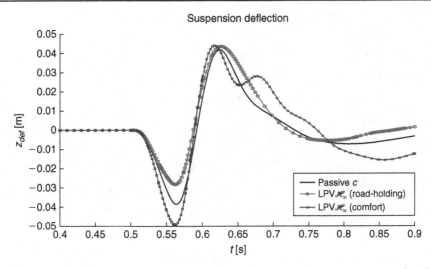

Figure 8.15: Bump test: Time response of the suspension deflection z_{def} – suspension limitations.

Figure 8.16 shows the responses of the unsprung mass displacement and of the tire deflection. In this case, the conclusions which can be drawn are consistent with the previous frequency-domain analysis:

- "LPV semi-active" Controller 1 parametrization provides bad damping, and filtering of the bump from the wheel point of view, while "LPV semi-active" Controller 2 damps the signal well.
- Improvements of tire deflection are not well demonstrated in this experiment but still Controller 2 limits tire deflection (it is even more clear in Figure 8.11 – bottom).

On Figure 8.17, the force vs. deflection speed diagram shows that the semi-active actuator limitations are fulfilled. Moreover, it confirms the previous FR plot, i.e.:

- "LPV semi-active" Controller 1 (comfort) parametrization mainly provides low damping factors.
- "LPV semi-active" Controller 2 (road-holding) parametrization mainly provides high damping factors.

These points are consistent with the idea of comfort and road-holding improvements.

8.7 Conclusions

In this chapter, a new strategy, called "LPV semi-active", that ensures the dissipative constraint for a semi-active suspension, while keeping the advantages and flexibility of the $\mathcal{H}_\infty$ control design is introduced. The main interests of such an approach are:

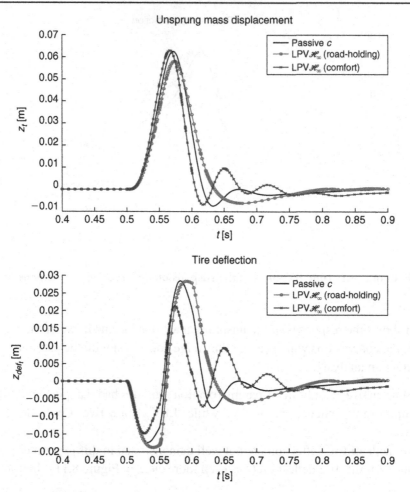

Figure 8.16: Bump test: Time response of the wheel displacement z_t (top) and the suspension deflection z_{def_t} (bottom) – road-holding criterion.

1. Flexible design (1): possibility to apply various approaches $\mathcal{H}_\infty$, $\mathcal{H}_2$, Pole placement, Mixed criteria, etc.

2. Flexible design (2): an infinite set of parametrization is possible (here two parametrizations are successfully illustrated to enhance either comfort or road-holding performances).

3. Measurement: only the suspension deflection (and its first derivative) is required.

4. Computation: synthesis leads to two LTI controllers and a simple scheduling strategy based on a static actuator model (no on-line optimization process involved).

5. Internal stability is preserved for all $\rho \in [\underline{\rho}; \overline{\rho}]$, thanks to the LPV $\mathcal{H}_\infty$ design.

6. Implementation: the solution is tractable for any kind of semi-active actuators.

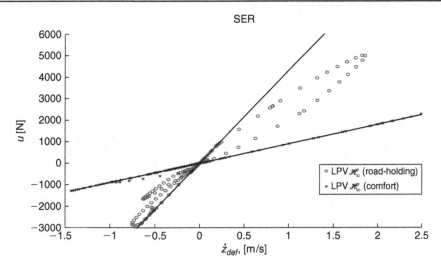

Figure 8.17: Bump test: Force vs. deflection speed diagram. $c_{min} = 900\,\text{Ns/m}$ and $c_{max} = 4300\,\text{Ns/m}$.

In the proposed "LPV semi-active" strategy, the semi-active constraint is handled through the LPV design. In this approach, the varying parameter is used to modify the control performance. The proposed controller shows good performance results through both frequency criteria and time simulation experiments, compared to well established existing semi-active controllers. This new "LPV semi-active" strategy exhibits significant improvements on the achieved and achievable performances and provides a non-negligible design flexibility.

Moreover, compared to LQ and MPC techniques, the implementation of such a controller results in a low cost solution in terms of controller order (resource consumption) and sensor requirements which is one of the key points in all embedded solutions. Compared to SH-ADD, it presents more flexibility but results in a more complex solution (from the computational point of view). The proposed approach can also be applied to any semi-active actuator based system.

CHAPTER 9

Conclusions and Outlook

The aim of this book was to discuss the semi-active suspension control problem for vehicles, following the roadmap presented in Figure 1.4. As indicated throughout the previous chapters, it is addressed to a large variety of readers, from undergraduate students (first chapters) to researchers and industrial experts in automotive, mechatronics and control theory (last chapters).

After introducing, in Chapters 2 and 3, the preliminary material to the non-familiar reader, namely, the semi-active technologies and the associated suspension-oriented vehicle models used for both the analysis and control synthesis purposes, Chapters 4 and 5 which play a pivotal role in the book, propose simple but efficient dedicated metric and methodology to evaluate suspension systems using time and frequency domain evaluation criteria. In the latter chapter, an innovative optimal performance bound is computed for both comfort and road-holding improvement objectives, in the semi-active case which the book focuses on. Then, Chapter 6 presents, in a condensed way, a state of the (exhaustive) art of the classical methodologies dedicated to the control of semi-active suspension systems. This chapter contains method presentation, bibliographical references and numerical simulations; it involves the criteria explained in Chapter 4, it gathers their major performance properties, and it allows a consistent classification and provides the reader with a reasonable overview of the classical methods. Finally, Chapters 7 and 8 propose a detailed analysis of two innovative semi-active suspension control strategies. The first one, called "mixed SH-ADD", definitely comfort oriented, takes its inspiration in mixing the complementary improvements of the Skyhook (SH) and the Acceleration Driven Damper (ADD) semi-active control methodologies to provide an optimal comfort filtering through a surprisingly simple and appealing control law. The second one, called "LPV semi-active", based on the recent developments of the linear control community (early 1990s) proposes, as the authors would admit, at the price of a heavier computation and modeling step, a very flexible solution that allows either comfort or road-holding enhancement. Both strategy performances greatly exceed the classical ones. Finally, Appendices A and B, following this conclusion, are more implementation oriented; indeed, Appendix A provides a comparison of the semi-active suspension control methods, taking into account the Flexibility/complexity trade-off, and emphasizing the cost complexity (in term of calculus, sensors and actuators). Then Appendix B describes a complete application, together with experimental results obtained on a motorcycle. This latter "chapter" illustrates some of the comfort oriented strategies, assessing the methodology presented throughout this book.

Nevertheless, it is not the authors' objective to recommend a strategy over another, because, as is (hopefully) emphasized throughout the book, semi-active suspension control is mainly a matter of compromise, and other recent methodologies (not deeply explored here) are still successfully applied to such systems (see Chapter 6) and clearly merit to be investigated further.

In the same way, comfort improvement has been a very active research field in the last decade and, naturally, a large variety of dedicated (and efficient) control strategies has emerged. As a consequence, one of the next issues in semi-active suspension control concerns the handling of stroke limitations which is linked to the comfort objective. Therefore, the development and implementation of road-holding oriented semi-active suspension strategies is now an open issue where no clear strategy has been successfully deployed.

Finally, the development of semi-active suspension control strategies is now strongly linked to the so-called global chassis control (GCC), and further issues will also concern the management of several control algorithms in a global vehicle dynamic framework.

Control Method Comparisons

In this chapter, an overview of the control methods presented in the book is given. These methods are analyzed through a comparative study, gathering both performance/flexibility and algorithm complexity. The underlying aim of this appendix is to help the reader to choose the appropriate control algorithm in view of an implementation on a practical semi-active system, and to provide a fair overview of the control methods including implementation complexity elements.

The chapter is simply organized as follows: Section A.1 summarizes the semi-active suspension control method properties and limitations and draws together all the control algorithms presented throughout the book. It summarizes limitations, algorithmic complexity, required actuators and sensors. Additionally, specific performances and flexibility are also discussed in order to provide the reader with a global view of the methodologies and orient him towards the appropriate solution to implement according to the objective and available resources (e.g. technological). Then, Section A.2 provides general conclusions. It also summarizes the common computational practical stiff point to be attentive to in view of an implementation.

A.1 Method Complexity Comparison

The aim of this section is to summarize the control methodology properties and emphasize the advantages and drawbacks of each approach from a performance, computational and solution cost point of view. The number of required sensors, algorithm complexity, implementation limitations and flexibility are summarized for each method. The aim is to provide a quick picture of the methodologies developed and give the reader a snapshot of a method allowing him to propose and evaluate his own semi-active control law with respect to the classical and innovative ones developed in this book.

To do so, a radar plot is introduced here to gather the performance flexibility and complexity of control algorithm information. This radar plot presents the following elements as extremities: Comfort/Road-holding/Simplicity, metrics between 0 and 1 (described thereafter).

- The "Comfort"/"Road-holding" extremities are computed using as reference result, the optimal bound computed in Chapter 5. Then, for each control algorithm, the following simple formulae will be used to set these two radar properties:

$$\text{Comfort} = \frac{2J_c^* - J_c}{J_c^*}$$

$$\text{Road-holding} = \frac{2J_{rh}^* - J_{rh}}{J_{rh}^*} \tag{A.1}$$

  where J_c and J_c^* (resp. J_{rh} and J_{rh}^*) denote the comfort (resp. road-holding) criterion of the considered control law and the optimal one respectively (see Chapter 5 and Definitions 4.1 and 4.2). Notice that when $J_c = J_c^*$ (resp. $J_{rh} = J_{rh}^*$) "Comfort" (resp. "road-holding") properties are equal to 1 which means that the control method is perfect for comfort (resp. road-holding) objective.

- The "Simplicity" criterion is given without any formal measurement, but according to the authors' feeling and background, considering the number of required sensors, control law complexity (multiplications and additions) and controller synthesis step complexity. This last information is not completely subjective, but still provides a classification of the method complexity. In a practical application, an automotive control engineer may clearly feel that a strategy is more appropriate than another according e.g. to the sensors achievable. A complex algorithm requiring a lot of measurements, which is time consuming and hard to tune will be rated as 0; while a simple one using few sensors, with neither specific tuning nor synthesis procedure will be rated at 1.

According to these descriptions, it is clear that a perfect algorithm should fill the entire radar plot (i.e. all the metrics to 1). Notice that, so far, as illustrated in this book, this is almost impossible due to the complementarity of the comfort and road-holding performances.

In the following, the presented algorithms are analyzed and (if possible) grouped.

A.1.1 Skyhook 2-States and Skyhook Linear

Sensors/Actuator From a sensor point of view, both methodologies require two sensors: one for the deflection velocity ($\dot{z}_{def}$) and one for the chassis velocity ($\dot{z}$). Since both algorithms use the sign of these sensors they should be precise enough. Concerning the required actuator, the 2-states version only requires a damper with a 2-states damper (minimal and maximal), while the linear version needs a continuously variable damper.

Implementation Complexity Both control laws are very simple and consist of a switching rule (logical condition). The 2-states strategy switches between minimal and maximal damping values; the linear version includes an additional parameter (α) and an apparently simple computation. Indeed, in this latter algorithm, a rational function term has to be computed (due to the division by $\dot{z}_{def}$) and may be complex for real time applications (or at least numerically

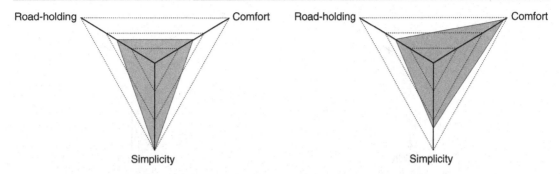

Figure A.1: Skyhook 2-states and linear performance/complexity radar diagram.

tricky). Finally, note that both approaches do not need any synthesis step. Consequently, these two algorithms may be considered as very simple and should be implemented at first on a practical test bench.

Algorithm Flexibility   Concerning the achieved performances, as illustrated in Chapter 6, both algorithms are clearly comfort oriented, but the algorithm flexibility is very poor. Indeed, the linear version is more flexible but still largely comfort oriented (see Figure A.1 and Chapter 6).

A.1.2 ADD and PDD

Sensors/Actuator   As the SH 2-states and the SH linear, the ADD and PDD semi-active suspension control strategies require two sensors:

- one for chassis acceleration ($\ddot{z}$) and one for deflection velocity ($\dot{z}_{def}$) for the ADD;
- one for deflection displacement (z_{def}) and one for velocity ($\dot{z}_{def}$) for the PDD control law.

Here again, since the algorithm is based on a switching rule according to the measurements, these latter must be accurate to provide accurate results. Regarding the required actuator, the ADD only requires a two-state damper, while the PDD requires a continuously variable damper.

Implementation Complexity   Considering the ADD control, the law only consists of a switching rule between minimal and maximal damping, while the PDD control consists in switching plus specific cases. Here again, no synthesis step is required, but the PDD algorithm needs additional computational operations that have to be solved on-line and since the suspension stiffness (k) is included in the control law calculus, a basic model knowledge is required as well.

Algorithm Flexibility   As with the SH 2-states approach, these two control algorithms do not present any tuning parameters and thus have frozen performance characteristics, i.e. only comfort. Flexibility is very poor, but the algorithm is more simple to implement (see Figure A.2 and Chapter 6).

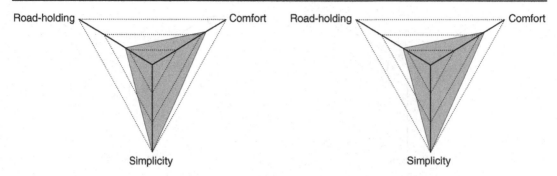

Figure A.2: ADD and PDD performance/complexity radar diagram.

A.1.3 Groundhook 2-States and Groundhook Linear

The Groundhook algorithms are hardly different from the above recalled Skyhook approaches. Characteristics are almost the same as those given in the SH subsection.

Sensors/Actuator　As with the SH 2-states, two sensors are required here. In the GH case, one sensor is dedicated to deflection velocity ($\dot{z}_{def}$) and the other for wheel velocity measurement ($\dot{z}_t$) (instead of chassis velocity in the SH cases). Concerning the actuator, the same comment as the SH holds: the 2-states version only requires a 2-states controllable damper while the linear version needs a continuously variable one.

Implementation Complexity　As for the SH, the Groundhook 2-states only consists in a switching rule between a minimal and maximal damping value, while the linear version presents an additional parameter (α) allowing us to adjust the performances. For both approaches, no synthesis step is required.

Algorithm Flexibility　As for the Skyhook 2-states (specialized in comfort), this algorithm is road-holding performances-oriented. Due to the additional parameter α the Groundhook linear version algorithm is road-holding oriented, but may be adjusted so that comfort performances are not diminished too much. It is interesting to keep in mind that, in spite of the apparent similarity/duality of the GH with respect to the SH, the GH algorithms do not provide road-holding performances as good as the SH algorithms in the comfort field (see Figure A.3 and Chapter 6).

A.1.4 SH-ADD (and 1 Sensor Version)

Sensors/Actuator　To apply this strategy, two measurements are required, the deflection velocity ($\dot{z}_{def}$) and the chassis acceleration ($\ddot{z}$) . The single sensor version (1 sensor) of this rationale allows us to use only chassis acceleration ($\ddot{z}$) and its integral. In this case, one single kind of sensor is needed, which is very interesting from an implementation point of view and solution cost as well. The actuator is simply a two-state one.

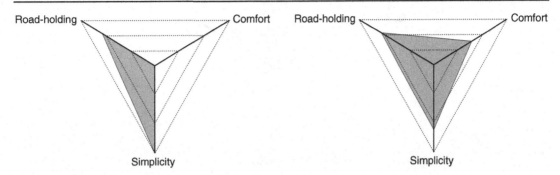

Figure A.3: Groundhook 2-states and linear performance/complexity radar diagram.

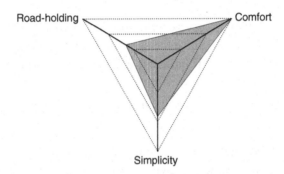

Figure A.4: SH-ADD performance/complexity radar diagram.

Implementation Complexity This algorithm is quite simple to implement since it only involves some conditional logic laws and one single parameter (α) to adjust the performance. Indeed, this parameter is very simple to adjust since it is directly related to the physical description of the vehicle (i.e. the first invariant point of transfer $F_z(s)$, around 20 rad/s in the motorcycle case used in this book, see ω_2 in Property 3.1 – Chapter 3). Note that one of the main complexities comes from the sign function which may be hard to compute on line, due to the noisy velocity measurements. To tackle velocity sign computation, it is better to use a hyperbolic tangent instead of the sign function. For this control, no synthesis is required.

Algorithm Flexibility This algorithm is clearly comfort oriented. The single tuning parameter allows us to select more sharply the frequency range selector (see Chapter 7). Therefore, this algorithm is poorly flexible but efficient for comfort purposes (see Figure A.4 and Chapter 6). For a deeper analysis of this control method, the reader is invited to refer to the experimental results summarized in Appendix B.

A.1.5 LPV Semi-Active

Sensors/Actuator For this strategy, as presented in this book (see Chapter 8), two measurements are required: the deflection displacement (z_{def}) and velocity ($\dot{z}_{def}$). But, as the

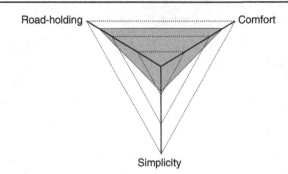

Figure A.5: LPV Semi-active linear performance/complexity radar diagram.

"LPV semi-active" methodology is very generic, for convenience, other measurements can be used (see e.g. Aubouet et al., 2009). Since the controller is designed using the $\mathcal{H}_\infty$ formalism, the choice of the used measurements is quite large. The actuator required has to be continuously variable.

Implementation Complexity   The complexity of this method comes from the synthesis and implementation:

- Assuming the model known, the weighting function definitions are, the authors must admit, far from trivial. Indeed, to design a controller using such an approach, the control engineer needs a model definition (i.e. including masses, suspension and tire spring stiffnesses) and should be able to design appropriate weighting filters before executing Algorithm 3 – Chapter 8, which is much more complex than previous methods. Of course Chapter 8 proposes parametrization examples to help the reader to achieve this first step.
- Then, the implementation of the controller is quite complex since it involves 2 state-space controllers and a (simple) scheduling strategy. The complexity of the implementation does not come from the LPV state-space formulation, which may be obtained using recent results (see Toth, 2008), but more from the gain values which are not "controlled".

Algorithm Flexibility   The counterpart of the algorithm complexity is that this approach is very flexible (see Chapter 8). In fact, it allows us to design controllers enhancing either comfort or road-holding performances. Moreover, it is to be noted that in the same way, it is possible to introduce an additional parameter allowing both comfort and road-holding by scheduling an external parameter and then to obtain comfort and road-holding performances as a function of the road situation (see Figure A.5 and Chapter 6).

A.1.6 (Hybrid) MPC Based
Sensors/Actuator   Concerning the MPC approaches, the full model state (x) measurement is required, which means for suspension control, at least the deflection displacement and velocity, and chassis displacement and velocity. Here again a continuously variable damper is

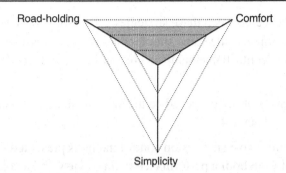

Figure A.6: (Hybrid) MPC performance/complexity radar diagram.

desired, but two-state may also be used, by including in the optimization problem formulation the discrete nature of the control signal (see Chapter 6).

Implementation Complexity   The synthesis step is quite simple, since it consists in formulating the problem as an "almost classical" optimization one (even if the criteria to minimize are not always so easy to define). Additionally, it is done in a discrete framework, which is interesting for implementation purposes. However, as the reader should probably have noticed, the main drawback is that this optimization problem has to be solved on-line to derive the optimal control law at each time step. Therefore, the MPC approaches are definitively time and energy consuming. In defence of these approaches, multi-parametric approaches (MPT, 2009) allow us to solve the problem off-line and then switch between subspaces, thus avoiding time consuming (without optimal guarantee) optimization.

Algorithm Flexibility   As for the "LPV semi-active", the counterpart of its complexity is the clear possibility to modify the performance objective, turning the suspension towards either comfort or road-holding objectives. Additionally, this problem formulation should also include static constraints such as deflection limitations, rate limitations, enhancing the possibility spectrum. In the authors' opinion, this approach is very promising but still very expensive in terms of consumption and may be hard to implement on a real vehicle (see Figure A.6 and Chapter 6; see also papers of Bemporad et al., 2003b). Indeed, an explicit formulation of the MPC exists, where the idea consists in defining controllers associated with zones, as a function of the state-space (see e.g. Bemporad et al., 2002).

A.2 Conclusions

It is important to note that the main difficulties for the simpler strategies, involving switching rules according to the sign of a given signal, is specifically the detection of the zero cross, which implies a good measurement of the considered signals. If this problem is overcome, the resulting control algorithm is very simple and these methods are greatly recommended for a first implementation. In fact, they provide good results in both comfort and road-holding.

Note that another difference between these strategies is that some of them require a continuously variable damper while some simpler ones only require a 2-states controllable damper. Practically a few controllable dampers are continuously variable, but may have a set of damping laws.

The achieved performances of the proposed controllers are analyzed thanks to the criteria and constraints presented in Chapter 4.

In this appendix, the semi-active suspension control methods presented in this book are reviewed and evaluated from both a performance point of view and a complexity one. The interest of the present appendix consists in providing the reader with an overview of the "cost" (e.g. sensor, actuator, algorithmic complexity) of each approach.

The present appendix may help the reader to choose which control strategy applies, taking into account the available sensors, actuators, and especially, calculus power and memory resources. Indeed, depending on the reader's semi-active suspension control and implementation maturity (and of course, time), different kinds of approaches should be tested.

In Appendix B, an illustration of the practical aspects and implementation procedure is sketched using some simple but efficient comfort-oriented control algorithms (e.g. SH, ADD and SH-ADD).

Case Study

The object of this Appendix is to present a complete case study of the design, implementation and testing of an electronic control system for a semi-active suspension. Every step is linked to a specific part of the book to give the reader a better understanding of the results given so far. The outline is as follows: in Section B.1, a description and characterization of the semi-active actuator is given. The quarter-car model for the rear suspension is recalled in Section B.2. Some of the control strategies already introduced in the Book are recalled in Section B.3. Section B.4 is devoted to the introduction of the experimental set-up and basics on the preprocessing of signals. The experimental protocol and results are illustrated in Sections B.5 and B.6 respectively.

B.1 Description of the Actuator

The semi-active shock absorber used in this work is a prototype damper installed on the rear axle of a hypersport-class motorcycle. This component is equipped with a current driven solenoid electrohydraulic valve; it can continuously change the damping ratio within its controllability range. These electrohydraulic valves have no embedded electronics; they must be commanded by an external electronic control unit (ECU), which implements a fast PI servo-loop having the goal of regulating the current at the desired value as described in Chapter 2. A concise model of the controllable shock absorber is given by the following equation (see Chapter 2):

$$\begin{cases} F_d(t, \dot{z}_{def}(t)) = c(t)\dot{z}_{def}(t) \\ \dot{c}(t) = -\beta c(t) + \beta c_{in}(t - \tau) \\ c_{min} \ll c_{in}(t) \ll c_{max} \end{cases} \tag{B.1}$$

Relation (B.1) represents a nonlinear multiple-inputs-single-output dynamical system. The inputs are the requested damping ratio c_{in} and the stroke speed of the shock absorber $\dot{z}_{def}$; the output is the damping force F_d. The dynamical behavior of the damping coefficient is described by a first-order differential equation with a delayed input τ and bandwidth represented by constant β. The force delivered by the shock absorber is proportional to the stroke speed, scaled by the actual damping ratio.

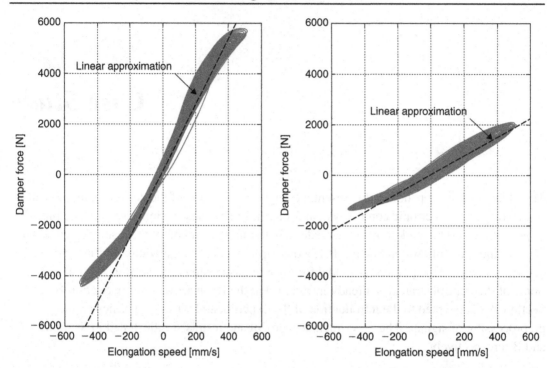

Figure B.1: Damper characteristics in the speed-force domain. Left: minimum damping c_{min}. Right: maximum damping c_{max}.

The damping request (proportional to the current in the solenoid valve) can vary continuously in the range between the minimum damping ratio c_{min} and the maximum damping ratio c_{max}.

The passive-like behavior of the damper is concisely illustrated in Figure B.1, where the characteristics of the damper are displayed in the classical speed-force domain. In detail, the response of the damper to a 10 Hz sinusoidal excitation, in its two extreme damping conditions ($c_{in} = c_{min}$ and $c_{in} = c_{max}$) is illustrated.

The ratio between the minimum and the maximum damping is the so-called controllability range of the shock absorber. Note that this ratio is about 1:3, and it is large enough to obtain good results with semi-active algorithms. On Figure B.1 the behavior of the maximum and minimum fixed damping can be observed and compared to the linear approximation as proposed by (B.1). Such an approximation is obtained with a linear regression of the experimental data. Note that the speed-force trajectories show a slight hysteresis and a little "regressive" behavior for very high elongation speed around ±400 mm/sec. This is mainly due to the asymmetric characteristics of the hydraulics circuits that manage the bound and rebound phase. In order to highlight the damping dynamics, the shock absorber can be tested by exciting the device with a constant elongation speed and, during excitation, by switching the damping ratio from minimum to maximum. Therefore, considering model (B.1) the damping

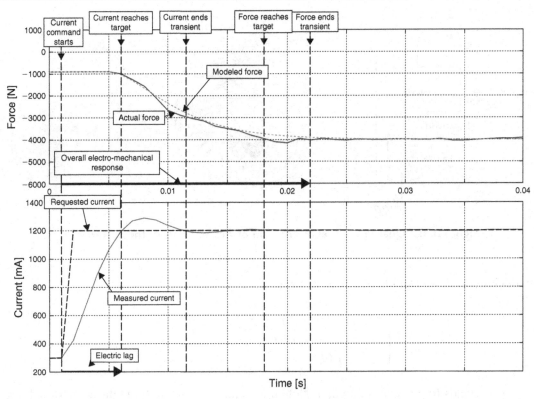

Figure B.2: **Details of the transient behavior of the damper subject to a step-like variation of the damping request.**

force is simply proportional to the damping variation. An example of this test is displayed in Figure B.2; in detail, it represents a switch during the compression phase (negative forces), when the damping request is changed from its minimum value c_{min} to its maximum value c_{max}.

Note that the force response to a step on the damping request shows approximately a linear 1st-order behavior, as assumed, although it looks slightly disturbed in the middle of the rising time. The valve current reaches the target in less than 10 ms, whereas the force reaches the target in less than 20 ms. Note that the current response is affected by a little overshoot. This is due to the ECU actuation which is limited to the voltage supply at maximum (for more details, see Chapter 2, where the technology is presented).

B.2 Model of the Semi-Active Suspension System

As depicted in Figure B.3, the rear suspension of the motorcycle equipped with a controllable damper may be modeled around its equilibrium point, according to the semi-active quarter-car

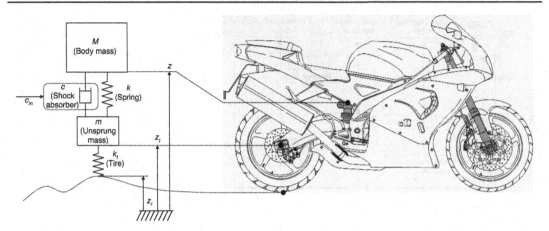

Figure B.3: "Quarter-car" representation of the rear part of the motorcycle.

model introduced in Definition 3.6, as follows:

$$\begin{cases} M\ddot{z} = -k(z - z_t) - c(\dot{z} - \dot{z}_t) \\ m\ddot{z}_t = k(z - z_t) + c(\dot{z} - \dot{z}_t) - k_t(z_t - z_r) \\ \dot{c} = \beta(c_{in} - c) \end{cases} \tag{B.2}$$

where c_{in} stands for the control input u. Note that, with reference to (B.1) the actuator delay τ has been ignored, since it is nested by the damper mechanical dynamic.

B.3 Control Algorithms

For control design purposes, we focus here on the minimization of the variation of the body vertical acceleration, namely (see also Section 4.1.1) :

$$\min \ddot{z}(t) = \min z(t) \tag{B.3}$$

This objective is the so-called "comfort-oriented" objective. Practically speaking this means that the main goal is to provide to the driver a high-quality filtering of the road disturbances, without deteriorating road-contact performance. Despite the "comfort" flavor, this objective is usually perceived in suspension control design, and also in high-performance sporty vehicles. In the following, the tested control strategies are recalled.

SH 2-State and SH Linear Controls   SH 2-state control is the semi-active heuristic approximation of the ideal concept of Skyhook damping; it is the most widely known control strategy in semi-active suspension systems. The two-state approximation of the Skyhook requires an ON–OFF controllable shock absorber; the control law is given by (see

Section 6.1.1.1):

$$c_{in} = \begin{cases} c_{min} & \text{if } \dot{z}\dot{z}_{def} \leq 0 \\ c_{max} & \text{if } \dot{z}\dot{z}_{def} > 0 \end{cases} \tag{B.4}$$

If a continuous modulation of the damping coefficient is available, a slightly more sophisticated expression of the SH algorithm, the SH linear, can be implemented (see Section 6.1.1.2):

$$c_{in} = \begin{cases} c_{min} & \text{if } \dot{z}\dot{z}_{def} \leq 0 \\ \text{sat}_{c_{in} \in [c_{min}; c_{max}]} \left(\dfrac{\alpha c_{max} \dot{z}_{def} + (1-\alpha) c_{max} \dot{z}}{\dot{z}_{def}} \right) & \text{if } \dot{z}\dot{z}_{def} > 0 \end{cases} \tag{B.5}$$

where α is a tuning parameter that affects the closed-loop performances. Specifically, if $\alpha = 1$ then the classical SH 2-states is obtained.

Acceleration Driven Damping (ADD) Control   The implementation of ADD control requires a two-level damper; the control law is given by (Section 6.1.2):

$$c_{in} = \begin{cases} c_{min} & \text{if } \ddot{z}\dot{z}_{def} \leq 0 \\ c_{max} & \text{if } \ddot{z}\dot{z}_{def} > 0 \end{cases} \tag{B.6}$$

The ADD control has been developed using optimal and nonlinear control theory; it has been proven to be the optimal control strategy when the goal objective is the minimization of the body vertical acceleration.

Mix-1-Sensor Control   Similarly to the SH and ADD, this strategy requires a two-level damper; the control law is given by (Section 7.2):

$$c_{in} = \begin{cases} c_{max} & \text{if } (\ddot{z}^2 - \alpha^2 \dot{z}^2) \leq 0 \\ c_{min} & \text{if } (\ddot{z}^2 - \alpha^2 \dot{z}^2) > 0 \end{cases} \tag{B.7}$$

This control law is extremely simple since, as the SH and ADD, it is based on a static rule which makes use of $\dot{z}$, $\ddot{z}$ only. Remember that the key idea of (B.7) is condensed in $(\ddot{z}^2 - \alpha^2 \dot{z})$, the frequency range selector.

Since this strategy computes only the amount $(\ddot{z}^2 - \alpha^2 \dot{z})$ it only requires the use of one single sensor (typically an accelerometer) for monitoring $\ddot{z}$; then by integration $\dot{z}$ can be derived. This feature makes the Mix-1-Sensor extremely appealing for real implementation (see also Chapter 6 and Appendix A).

B.4 Experimental Set-Up

The control algorithms are implemented on the following experimental layout, equipped on the vehicle:

Figure B.4: Example of sensor installation.

- The basic sensor set used is illustrated in Figure B.4. It is constituted by a body-side vertical MEMS accelerometer, having the range ± 10 g; a wheel-side vertical MEMS accelerometer, having the range ± 25 g; a stroke sensor, constituted by a potentiometer, having the range of 0–60 mm.
- The electronic control unit is based on the microcontroller devoted to automotive applications and equipped with a digital signal processing unit. The suspension code runs at a frequency of 1 kHz. The algorithms and filters are digitalized according to a Tustin transformation method.
- The shock absorber is driven by a feedback control of the current. The current is measured by a Hall-effect based transducer.

Overall, the algorithms presented so far indicate the following:

- The body vertical acceleration $\ddot{z}$.
- The body vertical velocity $\dot{z}$.
- The suspension stroke velocity $\dot{z}_{def}$.

In order to compute these algorithms, the body vertical acceleration is measured by the body accelerometer.

- Chassis velocity $\dot{z}$ is obtained by numerical derivation of $\ddot{z}$.
- Stroke velocity $\dot{z}_{def}$ is numerically derivated from the signal z_{def} measured by the potentiometer (alternatively it may be obtained by the difference of the integrated body and wheel acceleration).

The transfer function of the low-pass filter for numerical integration is:

$$I_n(s) = \frac{1}{s + \beta_i} \tag{B.8}$$

where β_i is the cut-off frequency of the filter. Practically speaking the integration is made over the bandwidth beyond the frequency β_i. Note that for relatively high frequencies the numerical integral is a good approximation of the ideal integration $I(s) = \frac{1}{s}$; indeed $I_n(j\omega) \approx \frac{1}{j\omega} \approx I(j\omega)$, if $\omega \gg \beta_i$ (where $s = j\omega$).

The use of $I_n(s)$ avoids the high gained noise at low frequency (known as drift) that affects ideal integration. A comparison of numerical and ideal integration is reported in Figure B.5, where the magnitude Bode diagrams of $I_n(s)$ and $I(s)$ are illustrated. For suspension design parameter β_i has been set at $\beta_i = 0.1 \cdot 2\pi$ (e.g. 0.1 Hz), so that the integration may be effective in the suspension bandwidth 1–25 Hz (typical of suspension dynamics).

Similar and dual conclusions may be drawn for derivation. The transfer function of the high-pass filter for numerical derivation is:

$$D_n(s) = \frac{\gamma s}{s + \gamma} \tag{B.9}$$

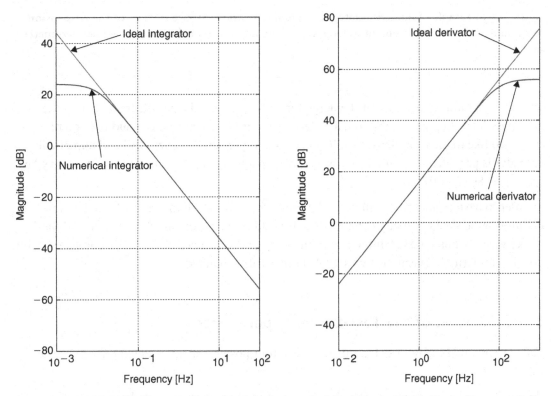

Figure B.5: Left: Bode diagram of the ideal and numerical integrator. **Right:** Bode diagram of the ideal and numerical derivator.

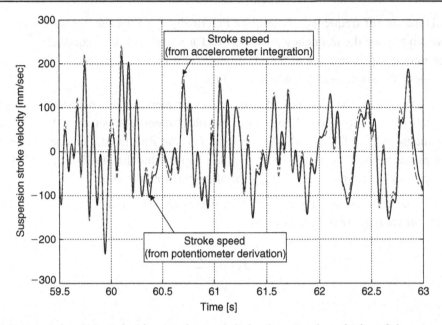

Figure B.6: Example of numerical integration and derivation. Stroke velocity of the suspension computed as derivation of potentiometer signal and difference of the body-wheel accelerometer signals.

where γ is a tuning parameter of the derivator and represents the cut-off frequency of the high-pass filter. In practice, the signal is derivated until frequency γ. Beyond that, no noise is amplified like in the ideal derivator $D(s) = s$. A comparison between ideal and numerical derivator is provided in Figure B.5. For suspension application the cut-off frequency may be set as $\gamma = 80 \cdot 2\pi$ (100 Hz).

In order to have a practical example of the presented method for integration and derivation, in Figure B.6 the stroke velocity computed as derivation of the potentiometer signal and the stroke velocity obtained for difference of the body and wheel acceleration are compared. Note the fair similarity between the two signals numerically obtained.

B.5 Definition of the Test-Bench Experiments

The experimental facility used in this work is a four poster test-rig, adapted for a two-wheeled vehicle.

Testing Protocol   The testing protocol has been defined in accordance with the evaluation methods introduced in Chapter 4. It is basically constituted by three types of experiments: the single-tone like experiments, the broad-band experiments, and the impulsive experiments.

- The first type of test-rig experiment is a time-varying sinusoidal excitation (usually nicknamed "frequency-sweep"). The explored frequency range is 0.1–30 Hz. The whole experiment lasts about 2 minutes, and the amplitude of the sinusoidal excitation decreases as the frequency increases. Note that this kind of excitation is the one defined in Algorithm 2, used to approximate Frequency Response; it allows a concise test of all the interested frequency range in a relatively short period of time. The test-rig displacement of this excitation is illustrated in the top frame of Figure B.7. This amplitude profile (as a function of the frequency) is designed in order to achieve the maximum amount of excitation, while avoiding the loss-of-contact of the wheel. In Figure B.7 an example of the signals measured on the vehicle (the two accelerations and the suspension stroke) when the damping ratio is set at its minimum value (c_{min}) is displayed. Notice that the occurrence of the two classical main resonances (the body resonance and the wheel resonance) is clearly shown by the stroke behavior (as discussed in the numerical analysis of the quarter-car model reported in Section 3.1.5). The filtering effect of the suspension is clearly visible by the direct comparison of the body and wheel accelerations.
- The second type of excitation is a broad-band signal, which reproduces a typical road profile. Such an excitation signal has been introduced in detail in Section 4.3.2 (Figure 4.8).
- The third type of excitation significantly differs from the sinusoidal sweep and the broad-band signal, since it is constituted by an impulsive-like excitation, such as the bump test (see also Chapter 4 and Figure 4.7).

Performance Metrics   The performance evaluation is carried out from three different perspectives, as extensively discussed in Chapter 4:

- Frequency-domain analysis, conducted according to the computation of the approximated transfer function from road profile to the body vertical acceleration (Algorithm 2, Section 4.2).
- Concise index analysis, based on the computation of Indexes in the time domain as introduced in Definition 4.2.
- Time domain performance evaluation stands for the direct analysis of the bump test (as discussed in Section 4.3).

B.6 Analysis of the Experimental Results

The reported results are displayed without scales for confidentiality reasons. However, note that since the analysis is based on relative comparisons, this has no influence on the conclusions that will be drawn.

In order to understand the basic behavior of the rear suspension of the vehicle, first it is interesting to analyze the performance of the vehicle without control algorithms (namely in a "traditional" configuration). In Figure B.8 the estimated frequency response from the road

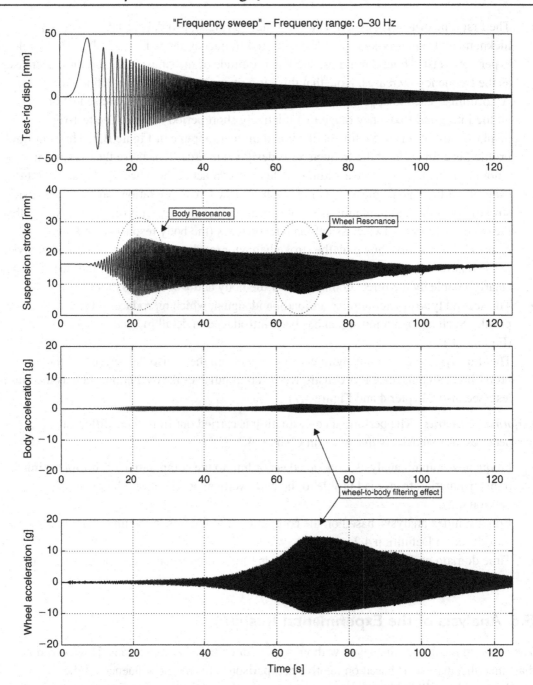

Figure B.7: Example of time-varying sinusoidal excitation experiment ("frequency sweep"), displayed in the time-domain.

acceleration to the body acceleration is displayed, when the damping radio is kept fixed at c_{min} and c_{max}. The obtained results are very clear and show the classical behavior of an under-damped and of an over-damped suspension:

- When c_{min} is set, the two main resonances are clearly visible since they are poorly damped (the body, around 2.5 Hz, and the wheel at 12 Hz).
- These resonances are fully damped when the damping ratio is set at its maximum value c_{max}. This better damping however is paid for in terms of poor filtering of the road disturbance beyond the body resonance. Given this trade-off, in a standard passive suspension an intermediate damping ratio is typically used, in order to find an acceptable compromise between resonance-damping and high-frequency filtering. This comment, now well known by the reader, confirms the simulation results provided in Chapters 3 and 4.

On Figure B.9, the (comfort) performance of the classical SH and ADD algorithms is analyzed, and is compared with the fixed-damping suspension. For the class of Skyhook algorithms, both the 2-state (labeled "SH-2-state") and SH linear or continuous (labeled

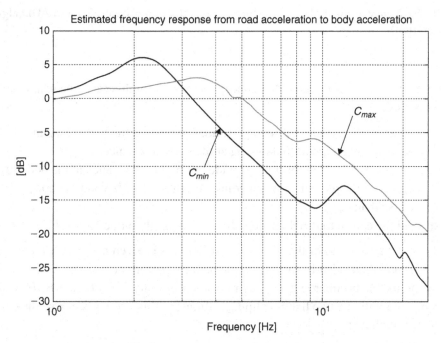

Figure B.8: Frequency domain filtering performance of the two extreme fixed damping ratios (sweep excitation).

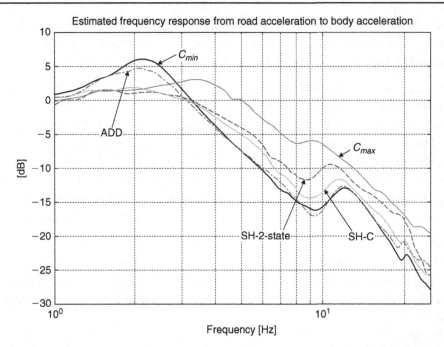

Figure B.9: Frequency domain filtering performance of the two classical SH and ADD algorithms (sweep excitation).

"SH-C") approximations of the ideal SH concept are tested. In this case, the results are easily interpreted as well:

- The ADD algorithm provides optimal performance beyond the body resonance (equal to c_{min}), while providing a medium damping at the body resonance.
- On the other hand, the SH algorithms provide optimal performance at the body resonance (equal to c_{max}), while guaranteeing a medium filtering effect beyond the body resonance (but still worse than c_{min}).
- As expected the continuous SH version algorithm slightly outperforms the 2-state one.

Overall, the best compromise of these classical algorithms is given by the continuous SH. The behavior of SH and ADD clearly shows complementary performances. The Mix-1-Sensor algorithm exploits this peculiar feature. Their behavior is analyzed in Figure B.10; as usual, they are compared with the two fixed-damping settings; in this comparison also the continuous SH algorithm is included.

Notice that the Mix SH-ADD 1-Sensor (labeled "Mix-1-S") algorithm shows a nearly-optimal comfort behavior, since it is able to stay on the lower bound of the c_{min}-c_{max} filtering curve (see Chapter 5 for the discussion on the optimal lower bound for suspension performance).

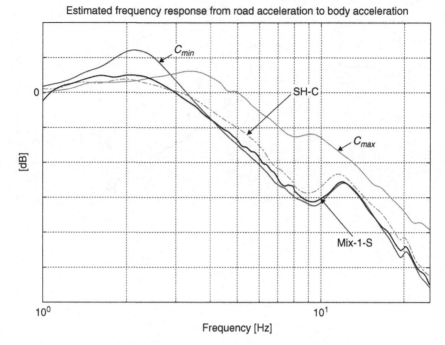

Figure B.10: Frequency domain filtering performance of the Mix-1-Sensor algorithm (sweep excitation).

In Figure B.11 the same frequency-domain analysis is presented, when the vehicle is excited with a random-walk signal. Notice that under the assumption of a genuine linear time-invariant system, the same results should be obtained whatever is the excitation. However, as long as the algorithms (B.4)–(B.7) are far from being linear, in principle, a different behavior could occur for different road profiles. The comparison of Figures B.10 and B.11 clearly reveals that the system is nonlinear; the main conclusions are comparable.

However, it is worth noticing that the difference between SH-C and Mix-1-S is less evident: SH-C is better at low frequency, whereas the Mix-1-S is better at mid and high frequency (beyond 3–4 Hz). In order to conclude, it is necessary to evaluate the suspension behaviors in the time domain also.

As always provided in this book, Figure B.12 gives a condensed view of the comfort performances of all the tested algorithms and fixed-damping configurations. The performance index used in Figure B.12 is a pure-comfort index defined as:

$$J_{\ddot{z}} = \frac{\mathscr{C}(\ddot{z}, \underline{t}, \bar{t})}{\mathscr{C}(\ddot{z}^{nom}, \underline{t}, \bar{t})} \tag{B.10}$$

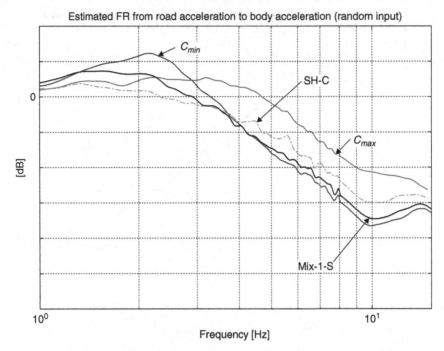

Figure B.11: **Frequency domain filtering performance of the SH and Mix-1-S algorithms (random walk excitation).**

where index $J_{\ddot{z}}$ indicates the ratio between the energy of the body acceleration ($\ddot{z}$) for a generic suspension system compared with the the energy of the body acceleration ($\ddot{z}^{nom}$) of a nominal suspension system. Note here that, in this case, the nominal damping has been favorably selected to be c_{max}, instead of a $c = 1500\,\text{Ns/m}$ in Definition 4.2.

Obviously, the smaller $J_{\ddot{z}}$ is, the better the filtering performances are. The bar plot of Figure B.12 has been obtained from the test-rig experiments with the random-walk excitation, and confirms the results drawn from the frequency-domain analysis (it also confirms the simulation results presented in this book).

It is interesting to observe that from Figure B.12 one could conclude that the Continuous SH and the fixed damping provide almost the same performance; however this condensed performance index hides the fact that SH provides the same filtering performance but with the advantage of a much better resonance damping; this fact can only be appreciated by the direct frequency-domain analysis.

We conclude this test-bench analysis by presenting an example of time-domain behavior of the suspension. In Figure B.13, the motorcycle response to a 45 mm bump is displayed. To make

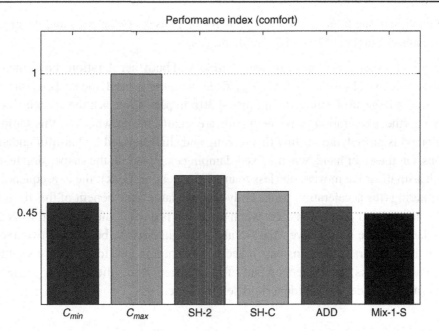

Figure B.12: Comparison of all the tested configurations using the condensed index J_c.

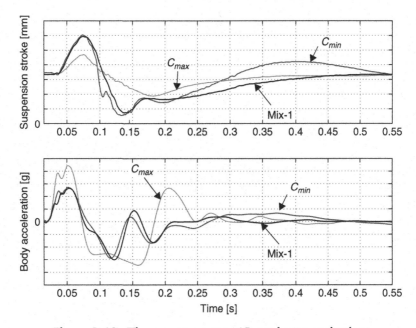

Figure B.13: Time response to a 45 mm bump excitation.

the interpretation of the figure more clear, only the responses of c_{min}, c_{max} and the best semi-active algorithm (the "Mix-1-S") are displayed.

The time-domain behavior of the suspension stroke and body acceleration shows the behavior of a good semi-active algorithm: when a c_{min} fixed damping is used, the suspension reacts to the bump with a large stroke movement (almost 30 mm peak-to-peak); the consequence is a good filtering (the acceleration peaks body-side are small); the drawback of this setting is that, when the bump is passed, the settling time is long and characterized by harmful undamped oscillations. On the other hand, when a fixed damping c_{max} is used, the suspension reacts to the bump with a small stroke movement (less than 15 mm peak-to-peak); the consequence of this is a poor filtering (the acceleration peaks body-side are large); the benefit of this fixed tuning is clear in the second part of the transient; when the bump is passed, the settling time is short and very well damped. The semi-active Mix-1-S algorithm inherits the best of the two fixed settings; in the first part of the transients it keeps the damping low, to obtain a good filtering (low acceleration peaks); in the second part of the transient it sets the damping to the maximum value, in order to provide a short settling time.

References

Ahmadian, M. and Simon, D. (2001). Vehicle evaluation of the performance of magneto rheological dampers for heavy truck suspensions. *Journal of Vibration and Acoustics*, 123:365–376.

Ahmadian, M. and Song, X. (1999). A non parametric model for magneto-rheological dampers. In *Proceedings of the ASME Design Engineering Technical Conference*, Las Vegas, Nevada, USA.

Ahmadian, M., Song, X., and Southward, S. (2004). No-jerk skyhook control methods for semiactive suspensions. *Transactions of the ASME*, 126:580–584.

Andreasson, J. and Bunte, T. (2006). Global chassis control based on inverse vehicle dynamics models. *Vehicle System Dynamics*, 44(supplement):321–328.

Apkarian, P. and Gahinet, P. (1995). A convex characterization of gain scheduled $\mathscr{H}_\infty$ controllers. *IEEE Transaction on Automatic Control*, 40(5):853–864.

Årzén, K.-E. (2003). *Real-Time control systems*. Lecture note of the Lund Institute of Technology.

Åström, K.-J. and Wittenmark, B. (1997). *Computer controlled systems: Theory and design*, 3rd Edition. Prentice Hall.

Aubouet, S., Sename, O., Dugard, L., Poussot-Vassal, C., and Talon, B. (2009). Semi-active $\mathscr{H}_\infty$/LPV control for an industrial hydraulic damper. In *9th European Control Conference (ECC)*, Budapest, Hungary.

Aubouet, S., Sename, O., Talon, B., Poussot-Vassal, C., and Dugard, L. (2008). Performance analysis and simulation of a new industrial semi-active damper. In *Proceedings of the 17th IFAC World Congress (WC)*, Seoul, South Korea.

Balas, G. J., Bokor, J., and Szabo, Z. (2003). Invariant subspaces for LPV systems and their application. *IEEE Transaction on Automatic Control*, 48(11):2065–2069.

Bemporad, A., Borrelli, F., and Morari, M. (2002). Model predictive control based on linear programming – The explicit solution. *IEEE Transaction on Automatic Control*, 47(12):1974–1985.

Bemporad, A., Borrelli, F., and Morari, M. (2003a). Min-Max control of constrained uncertain discrete-time linear systems. *IEEE Transaction on Automatic Control*, 48(9):1600–1606.

Bemporad, A., Morari, M., Dua, V., and Pistikopoulos, E. (2003b). The explicit linear quadratic regulator for constrained systems. *Automatica*, 38(1):3–20.

Biannic, J.-M. (1996). *Robust control of parameter varying systems: aerospace applications*. Ph.D. thesis (in French), Université Paul Sabatier – ONERA, Toulouse, France.

Borrelli, F., Baotic, M., Bemporad, A., and Morari, M. (2003). An efficient algorithm for computing the state feedback optimal control law for discrete time hybrid systems. In *Proceedings of the IEEE American Control Conference (ACC)*, Denver, Colorado, USA.

Bosch (2000). *Automotive Handbook*, 5th Edition. Bosch Gmbh.

Boyd, S., El-Ghaoui, L., Feron, E., and Balakrishnan, V. (1994). *Linear Matrix Inequalities in System and Control Theory*. SIAM Studies in Applied Mathematics.

Briat, C. (2008). *Robust Control and Observation of LPV Time-Delay Systems*. Ph.D. thesis, Grenoble INP, GIPSA-lab, Control Systems Dept., Grenoble, France.

Bruzelius, F. (2004). *Linear Parameter-Varying Systems: an approach to gain scheduling*. Ph.D. thesis, University of Technology of Göteborg, Sweden.

Burckhardt, M. (1993). *Fahrwerktechnik: Radschlupf-Regelsysteme*. Vogel-Verlag.

Canale, M., Milanese, M., and Novara, C. (2006). Semi-active suspension control using fast model-predictive techniques. *IEEE Transaction on Control System Technology*, 14(6):1034–1046.

Canudas, C., Velenis, E., Tsiotras, P., and Gissinger, G. (2003). Dynamic tire friction models for road/tire longitudinal interaction. *Vehicle System Dynamics*, 39(3):189–226.

Chilali, M., Gahinet, P., and Apkarian, P. (1999). Robust pole placement in LMI regions. *IEEE Transaction on Automatic Control*, 44(12):2257–2270.

Choi, S., Nam, M., and Lee, B. (2000). Vibration control of a MR seat damper for commercial vehicles. *Journal of Intelligent Material Systems and Structures*, 11:936–944.

Chou, H. and d'Andréa-Novel, B. (2005). Global vehicle control using differential braking torques and active suspension forces. *Vehicle System Dynamics*, 43(4):261–284.

Codeca, F., Savaresi, S., Spelta, C., Montiglio, M., and Ieluzzi, M. (2007). Semiactive control of a secondary train suspension. In *IEEE/ASME International Conference on Advanced Intelligent Mechatronics*, Zurich, Switzerland.

Corno, M. (2009). *Active stability control design for two-wheeled vehicles*. Ph.D. thesis, Politecnico di Milano, dipartimento di Elettronica e Informazione, Milano, Italy.

Delphi (2008). Delphi website. Technical report, Delphi. http://www.delphi.com.

Deprez, K., Moshou, D., Anthonis, J., Baerdemaeker, J. D., and Ramon, H. (2005). Improvement of vibrational comfort on agricultural vehicles by passive and semi-active cabin suspensions. *Computers and Electronics in Agriculture*, 49:431–440.

Di-Cairano, S., Bemporad, A., Kolmanovsky, I., and Hrovat, D. (2007). Model predictive control of magnetically actuated mass spring dampers for automotive applications. *International Journal of Control*, 80(11):1701–1716.

Dixon, J. (2007). *Shock Absorber Handbook*. Wiley.

Dorf, R. and Bishop, R. (2001). *Modern control systems*, volume 9th Edition. Prentice Hall.

Dorling, R., Smith, M., and Cebon, D. (1995). Achievable dynamic response of active suspensions in bounce and roll. In *IFAC Workshop on Advances in Automotive Control*, pages 63–70, Monte Verit, Switzerland.

Du, H., Sze, K., and Lam, J. (2005a). Semi-active $\mathcal{H}_\infty$ control with magneto-rheological dampers. *Journal of Sound and Vibration*, 283(3–5):981–996.

Du, H., Sze, K. Y., and Lam, J. (2005b). Semi-active $\mathcal{H}_\infty$ control with magneto-rheological dampers. *Journal of Sound and Vibration*, 283(3–5):981–996.

Emura, J., Kakizaki, S., Yamaoka, F., and Nakamura, M. (1994). Development on the semi-active suspension system based on the sky-hook damper theory. *Society of Automotive Engineers*, pages 17–26.

Fischer, D. and Isermann, R. (2003). Mechatronic semi-active and active vehicle suspensions. *Control Engineering Practice*, 12(11):1353–1367.

Flores, L., Drivet, A., Ramirez-Mendoza, R., Sename, O., Poussot-Vassal, C., and Dugard, L. (2006). Hybrid optimal control for semi-active suspension systems. In *Proceedings of the 10th Mini Conference on Vehicle System Dynamics, Identification and Anomalies*, Budapest, Hungary.

Gáspár, P. and Bokor, J. (2006). A fault-tolerant rollover prevention system based on LPV method. *International Journal of Vehicle Design*, 42(3–4):392–412.

Gáspár, P., Szabó, Z., Bokor, J., Poussot-Vassal, C., Sename, O., and Dugard, L. (2007). Toward global chassis control by integrating the brake and suspension systems. In *Proceedings of the 5th IFAC Symposium on Advances in Automotive Control (AAC)*, Aptos, California, USA.

Gáspár, P., Szaszi, I., and Bokor, J. (2004a). Active suspension design using LPV control. In *Proceedings of the 1st IFAC Symposium on Advances in Automotive Control (AAC)*, pages 584–589, Salerno, Italy.

Gáspár, P., Szaszi, I., and Bokor, J. (2004b). The design of a combined control structure to prevent the rollover of heavy vehicles. *European Journal of Control*, 10(2):148–162.

Gáspár, P., Szaszi, I., and Bokor, J. (2004c). Rollover stability control for heavy vehicles by using LPV model. In *Proceedings of the 1st IFAC Symposium on Advances in Automotive Control (AAC)*, Salerno, Italy.

Gáspár, P., Szaszi, I., and Bokor, J. (2005). Reconfigurable control structure to prevent the rollover of heavy vehicles. *Control Engineering Practice*, 13(6):699–711.

Gillespie, T. (1992). *Fundamental of vehicle dynamics*. Society of Automotive Engineers.

Giorgetti, N., Bemporad, A., Tseng, H., and Hrovat, D. (2005). Hybrid model predictive control application towards optimal semi-active suspension. In *Proceedings of the IEEE International Symposium on Industrial Electronics (ISIE)*, pages 391–397, Dubrovnik, Croatia.

Giorgetti, N., Bemporad, A., Tseng, H., and Hrovat, D. (2006). Hybrid model predictive control application toward optimal semi-active suspension. *International Journal of Control*, 79(5):521–533.

Girardin, G., Peter, T., Gissinger, G., and Basset, M. (2006). Modélisation non linéaire du confort dynamique d'un véhicule. In *Proceedings of the 17th Conférence Internationale Francophone d'Automatique (CIFA)*, Bordeaux, France.

Giua, A., Melas, M., Seatzu, C., and Usai, G. (2004). Design of a predictive semiactive suspension system. *Vehicle System Dynamics*, 41(4):277–300.

GLPK (2009). GLPK – GNU Linear Programming Kit.

Goodall, R. and Kortum, W. (2002). Mechatronic developments for railway vehicles of the future. *Control Engineering Practice*, 10(8):887–898.

Guglielmino, E. and Edge, K. (2004). Controlled friction damper for vehicle applications. *Control Engineering Practice*, 12(4):431–443.

Guglielmino, E., Sireteanu, T., Stammers, C., Ghita, G., and Giuclea, M. (2008). *Semi-active Suspension Control: Improved Vehicle Ride and Road Friendliness*. Springer, London.

Guglielmino, E., Stammers, C., Stancioiu, D., and Sireteanu, T. (2005). Conventional and nonconventional smart damping systems. *International Journal of Vehicle Autonomous Systems*, 3(2–4):216–229.

Hong, K.-S., Sohn, H.-C., and Hedrick, J.-K. (2002). Modified skyhook control of semi-active suspensions: A new model, gain scheduling, and hardware-in-the-loop tuning. *ASME Journal of Dynamic Systems, Measurement, and Control*, 124(1):158–167.

Hrovat, D. (1997). Survey of advanced suspension developments and related optimal control application. *Automatica*, 33(10):1781–1817.

Ieluzzi, M., Turco, P., and Montiglio, M. (2006). Development of a heavy truck semi-active suspension control. *Control Engineering Practice*, 14(3):305–312.

Isermann, R. (2003). *Mechatronic Systems: Fundamentals*. Springer-Verlag.

ISO2631 (2003). *ISO 2631: Mechanical vibration and shock – Evaluation of human exposure to whole-body vibration*. International Organization for Standardization.

Karnopp, D. (1983). Active damping in road vehicle suspension systems. *Vehicle System Dynamics*, 12(6):296–316.

Karnopp, D., Crosby, M., and Harwood, R. (1974). Vibration control using semi-active force generators. *Journal of Engineering for Industry*, 96(2):619–626.

Kawabe, T., Isobe, O., Watanabe, Y., Hanba, S., and Miyasato, Y. (1998). New semi-active suspension controller design using quasi-linearization and frequency shaping. *Control Engineering Practice*, 6(10):1183–1191.

Kiencke, U. and Nielsen, L. (2000). *Automotive Control Systems*. Springer-Verlag.

Koo, J. (2003). *Using Magneto-Rheological Dampers in Semiactive Tuned Vibration Absorbers to Control Structural Vibrations*. Ph.D. thesis, Virginia Polytechnic Institute and State University.

Koo, J. H., Ahmadian, M., Setareh, M., and Murray, T. (2004a). In search of suitable control methods for semi-active tuned vibration absorbers. *Journal of Vibration and Control*, 10(2):163–174.

Koo, J.-H., Goncalves, F., and Ahmadian, M. (2004b). Investigation of the response time of magnetorheological fluid dampers. *SPIE*, 5386:63–71.

Lofberg, J. (2004). YALMIP: A toolbox for modeling and optimization in MATLAB. In *Proceedings of the CACSD Conference*, Taipei, Taiwan.

Lord (2008). Lord website. Technical report, Lord. http://www.lord.com.

Margolis, D. L. (1983). Semi-active control of wheel hop in ground vehicles. *Vehicle System Dynamics*, 12(6):317–330.

Milliken, W. and Milliken, D. (1995). *Race car vehicle dynamics*. Society of Automotive Engineers.

Moreau, X. (1995). *La dérivation non entiere en isolation vibratoire et son application dans le domaine de l'automobile. La suspension CRONE: du concept à la réalisation*. Ph.D. thesis (in French), Université de Bordeaux I.

Morselli, R. and Zanasi, R. (2008). Control of a port hamiltonian systems by dissipative devices and its application to improve the semi-active suspension behavior. *Mechatronics*, 18(7):364–369.

MPT (2009). MPT – Multi Parametric Toolbox.

Niculescu, S-I. (2001). *Delay effects on stability. A robust control approach*, volume 269. Springer-Verlag: Heidelberg.

Oustaloup, A., Moreau, X., and Nouillant, M. (1996). The CRONE suspension. *Control Engineering Practice*, 4(8):1101–1108.

Poussot-Vassal, C. (2007). Discussion paper on: "Combining Slip and Deceleration Control for Brake-by-Wire Control Systems: a Sliding-Mode Approach". *European Journal of Control*, 13(6):612–615.

Poussot-Vassal, C. (2008). *Robust Multivariable Linear Parameter Varying Automotive Global Chassis Control*. Ph.D. thesis (in English), Grenoble INP, GIPSA-lab, Control System Dept., Grenoble, France.

Poussot-Vassal, C., Sename, O., and Dugard, L. (2009). Attitude and handling improvements based on optimal skyhook and feedforward strategy with semi-active suspensions. *International Journal of Vehicle Autonomous Systems: Special Issue on Modelling and Simulation of Complex Mechatronic Systems*, 6(3–4):308–329.

Poussot-Vassal, C., Sename, O., Dugard, L., Gáspár, P., Szabó, Z., and Bokor, J. (2006). Multi-objective qLPV $\mathcal{H}_\infty/\mathcal{H}_2$ control of a half vehicle. In *Proceedings of the 10th Mini-conference on Vehicle System Dynamics, Identification and Anomalies (VSDIA)*, Budapest, Hungary.

Poussot-Vassal, C., Sename, O., Dugard, L., Gáspár, P., Szabó, Z., and Bokor, J. (2007). A LPV based semi-active suspension control strategy. In *Proceedings of the 3rd IFAC Symposium on System Structure and Control (SSSC)*, Iguacu, Brazil.

Poussot-Vassal, C., Sename, O., Dugard, L., Gáspár, P., Szabó, Z., and Bokor, J. (2008a). Attitude and handling improvements through gain-scheduled suspensions and brakes control. In *Proceedings of the 17th IFAC World Congress (WC)*, Seoul, South Korea.

Poussot-Vassal, C., Sename, O., Dugard, L., Gáspár, P., Szabó, Z., and Bokor, J. (2008b). The design of a chassis system based on multi-objective qLPV control. *Periodica Polytechnica*, 36(1–2):93–97.

Poussot-Vassal, C., Sename, O., Dugard, L., Gáspár, P., Szabó, Z., and Bokor, J. (2008c). A New semi-active suspension control strategy through LPV technique. *Control Engineering Practice*, 16(12):1519–1534.

Ramirez-Mendoza, R. (1997). *Sur la modélisation et la commande de véhicules automobiles*. Ph.D. thesis (in French), INPG, Laboratoire d'Automatique de Grenoble (now GIPSA-lab), Grenoble, France.

Roos, C. (2007). *Contribution to the control of saturated systems in the presence of uncertainties and parametric variations: Application to on-ground aircraft control*. Ph.D. thesis (in French), Université Paul Sabatier – (ONERA-DCSD), Toulouse, France.

Rossi, C. and Lucente, G. (2004). $\mathcal{H}_\infty$ control of automotive semi-active suspensions. In *Proceedings of the 1st IFAC Symposium on Advances in Automotive Control (AAC)*, Salerno, Italy.

Sachs (2008). Sachs website. Technical report, Sachs. http://www.zfsachs.com.

Sammier, D. (2001). *Sur la modélisation et la commande des suspensions automobiles*. Ph.D. thesis (in French), INPG, Laboratoire d'Automatique de Grenoble (now GIPSA-lab), Grenoble, France.

Sammier, D., Sename, O., and Dugard, L. (2000). $\mathcal{H}_\infty$ control of active vehicle suspensions. In *Proceedings of the IEEE International Conference on Control Applications (CCA)*, pages 976–981, Anchorage, Alaska.

Sammier, D., Sename, O., and Dugard, L. (2003). Skyhook and $\mathcal{H}_\infty$ control of active vehicle suspensions: some practical aspects. *Vehicle System Dynamics*, 39(4):279–308.

Sampson, D. and Cebon, D. (2003). Active roll control of single unit heavy road vehicles. *Vehicle System Dynamics*, 40(4):229–270.

Savaresi, S., Bittanti, S., and Montiglio, M. (2005a). Identification of semi-physical and black-box non-linear models: the case of MR-dampers for vehicles control. *Automatica*, 41:113–117.

Savaresi, S., Siciliani, E., and Bittanti, S. (2005b). Acceleration driven damper (ADD): an optimal control algorithm for comfort oriented semi-active suspensions. *ASME Transactions: Journal of Dynamic Systems, Measurements and Control*, 127(2):218–229.

Savaresi, S., Silani, E., and Bittanti, S. (2004). Semi-active suspensions: an optimal control strategy for a quarter-car model. In *Proceedings of the 1st IFAC Symposium on Advances in Automotive Control (AAC)*, pages 572–577, Salerno, Italy.

Savaresi, S. and Spelta, C. (2007). Mixed sky-hook and ADD: Approaching the filtering limits of a semi-active suspension. *ASME Transactions: Journal of Dynamic Systems, Measurement and Control*, 129(4):382–392.

Savaresi, S. and Spelta, C. (2009). A single sensor control strategy for semi-active suspension. *IEEE Transaction on Control System Technology*, 17(1):143–152.

Savaresi, S., Spelta, C., Moneta, A., Tosi, F., Fabbri, L., and Nardo, L. (2008). Semi-active control strategies for high-performance motorcycles. In *Proceedings of the 2008 IFAC World Congress*, pages 4689–4694, Seoul, South Korea.

Scherer, C., Gahinet, P., and Chilali, M. (1997). Multiobjective output-feedback control via LMI optimization. *IEEE Transaction on Automatic Control*, 42(7):896–911.

Sename, O. and Dugard, L. (2003). Robust $\mathcal{H}_\infty$ control of quarter-car semi-active suspensions. In *Proceedings of the European Control Conference (ECC)*, Cambridge, England.

Shamma, J. and Athans, M. (1991). Guaranteed properties of linear parameter varying gain scheduled control systems. *Automatica*, 27(3):559–564.

Shamma, J. and Athans, M. (1992). Gain scheduling: Possible hazards and potential remedies. *IEEE Control Systems Magazine*, pages 101–107.

Shuqui, G., Shaopu, Y., and Cunzgi, P. (2006). Dynamic modeling of magnetorheological damper behaviors. *Journal of Intelligent Material Systems And Structures*, 17:3–14.

Simon, D. (2001). *An Investigation of the Effectiveness of Skyhook Suspensions for Controlling Roll Dynamics of Sport Utility Vehicles Using Magneto-Rheological Dampers*. Ph.D. thesis, Virginia Polytechnic Institute and State University.

Sohn, H., Hong, K., and Hedrick, J. (2000). Semi-active control of the Macpherson suspension system: Harware-in-the-loop simulations. In *IEEE CCA 2000*, pages 982–987, Anchorage, Alaska.

Song, X., Ahmadian, M., and Southward, S. (2007). Analysis and strategy for superharmonics with semiactive suspension control systems. *ASME Journal of Dynamic Systems, Measurement, and Control*, 129(6):795–803.

Spelta, C. (2008). *Design and applications of semi-active suspension control systems*. Ph.D. thesis, Politecnico di Milano, dipartimento di Elettronica e Informazione, Milano, Italy.

Spelta, C., Cutini, M., Bertinotti, S., Savaresi, S., Previdi, F., Bisaglia, C., and Bolzern, P. (2009). A new concept of semi-active suspension with controllable damper and spring. In *Proceeding of the European Control Conference 2009*, pages 4410–4415, Budapest, Hungary.

Spelta, C., Savaresi, S., and Fabbr, L. (2010). Experimental analysis and development of a motorcycle semi-active 1-sensor rear suspension. *Control Engineering Practice* (in press). Doi: 10.1016/j.conengprac.2010.02.006.

Spencer, B., Dyke, S., Sain, M., and Carlson, J. (1997). Phenomenological model of magnetorheological damper. *Journal of Engineering Mechanics*, 123:230–238.

Sturm, J. F. (1999). Using SeDuMi 1.02, a MATLAB toolbox for optimization over symmetric cones. *Optim. Methods Softw.*, 11/12(1–4):625–653. Interior point methods.

Tanelli, M. (2007). *Active Braking Control Systems Design for Road Vehicles*. Ph.D. thesis, Politecnico di Milano, dipartimento di Elettronica e Informazione, Milano, Italy.

Tondel, P., Johansen, T., and Bemporad, A. (2003). An algorithm for multi-parametric quadratic programming and explicit MPC solutions. *Automatica*, 39(3):489–497.

Toth, R. (2008). *Modeling and identification of linear parameter-varying systems. An orthogonal basis function approach*. Ph.D. thesis, DISC, Delft, Netherland.

Tseng, H. and Hedrick, J. (1994). Semi-active control laws – optimal and sub-optimal. *Vehicle System Dynamics*, 23(1):545–569.

Valasek, M. and Kortum, W. (2002). *The Mechanical Systems Design Handbook*, chapter on Semi-Active Suspension Systems II. CRC Press LLC.

Velenis, E., Tsiotras, P., Canudas, C., and Sorine, M. (2005). Dynamic tire friction models for combined longitudinal and lateral vehicle motion. *Vehicle System Dynamics*, 43(1):3–29.

Zhou, K., Doyle, J., and Glover, K. (1996). *Robust and Optimal Control*. Prentice Hall.

Zin, A. (2005). *Robust automotive suspension control toward global chassis control*. Ph.D. thesis (in French), INPG, Laboratoire d'Automatique de Grenoble (now GIPSA-lab), Grenoble, France.

Zin, A., Sename, O., Gaspar, P., Dugard, L., and Bokor, J. (2008). Robust LPV – $\mathcal{H}_\infty$ control for active suspensions with performance adaptation in view of global chassis control. *Vehicle System Dynamics*, 46(10):889–912.

Index

Printed in the United States
By Bookmasters